BRAVE NEW JUDAISM

BRAVE NEW JUDAISM

When Science and Scripture Collide

MIRYAM Z. WAHRMAN

Brandeis University Press

Published by University Press of New England

Hanover and London

Brandeis University Press

Published by University Press of New England,

One Court St., Lebanon, NH 03766

© 2002 by Brandeis University Press

All rights reserved

Printed in the United States of America

5 4 3 2 1

Library of Congress Cataloging-in-Publication Data

Wahrman, Miryam Z.

Brave new Judaism : when science and scripture collide / Miryam Z.
Wahrman.

p.   cm.

Includes bibliographical references and index.

ISBN 1–58465–031–1

1. Judaism and science. 2. Bioethics—Religious aspects—Judaism.
3. Cloning—Religious aspects—Judaism. 4. Medicine—Religious
aspects—Judaism. 5. Technology and Jewish law. I. Title.

BM538.S3 W34 2002

296.3'4957—dc21   2002004930

With deepest love and gratitude to:
Israel, Abby, and Susie Wahrman
and
Zev Zahavy

And in loving memory of:
Edith Zahavy, z"l

Contents

Preface

Staying Alive

They say there are no atheists in foxholes. I say there are no atheists in ICUs. Not even the doctors . . . especially not the doctors.

"Your mother's getting better." Those four words meant the world to me after my mother had gone through her fifth serious operation in a span of two weeks. When my mother was seriously ill, I learned for the first time how strong the will to live can be. Mom went through operation after operation, and kept fighting back. And watching my mother struggle to breathe, to eat, to communicate, made me want to fight as hard as I could to get the medical care needed to keep her alive—and to help her get better. Her will to live was so strong that she was able to overcome many challenges: an endartectomy (clearing out the artery to the brain), a quintuple bypass, a pacemaker, two major surgeries on her intestines, plastic surgery on her chest muscles. Unable to speak because of a tracheostomy, she would hold up her fingers counting the surgeries—four, five, six . . .

"They perform miracles here every day," asserted my brother, after my mother pulled through another operation. And, indeed, it seemed as if medical miracles would lead to her recovery. For month after month in the intensive care unit doctors and nurses shook their heads in wonderment as she recovered time after time. Sophisticated medical science and technology kept her alive for month after month.

And when my mother died, succumbing to an infection after her almost six-month struggle, it was difficult to ascertain for sure when life had ended, because of all the technology keeping her alive. As I watched the heart monitor, the peaks and valleys on the screen were replaced by a flat line, broken by an occasional blip. Even in death, the pacemaker continued its work, futilely attempting to regulate a no longer functioning heart. Her lungs continued to inflate and deflate, due to the ventilator attached to her throat (that machine had kept her alive for more than five months). The whole scene was surreal, sinking into my consciousness only slowly. It was part daytime soap opera and part confusing and crushing tragedy. Moments later when my father walked into the hospital room he asked, "but how is she gone, if she's still breathing?" Her heart had indeed stopped even though the machines kept going. The machines attached to her lifeless body were finally turned off—and they stopped their beeping and humming. I stroked her hand and gently closed her eyes.

The drive to stay alive may be partially intellectual and rational. But mostly it's emotional, and sometimes even irrational. This drive has helped people survive life-threatening crises throughout the ages. And this drive to stay alive, or to protect a loved one from death, has led to many conundrums in the application of modern biomedical science.

The Drive to Reproduce

On the other side of the coin is the irrational, emotional, almost irresistible urge to reproduce. The most obvious evidence of this drive is the significant role sex plays in human life. Another manifestation of this drive to reproduce is the ticking of a so-called "biological clock" in older women. The desire to reproduce can lead to feelings of desperation in childless couples who have experienced years of infertility. The need and desire to conceive has also led to the development of new technologies and ultimately, to many bioethical dilemmas.

As a childless couple in the 1980s, my husband and I were privy to all the latest developments in infertility research. The year 1978 had seen the birth of the first test-tube baby, Louise Brown. In the early 80s I was working in that fledgling field and was involved in setting up the first in vitro fertilization (IVF) lab in New York. Our IVF team succeeded in producing the first test-tube baby in the state (born in 1984). It was an exciting and heady time—it seemed as if the team were producing miracle babies, and at the same time we were challenged with a multitude of ethical questions that were novel and daunting. Ironically, although I was working in the field, the feat of conception eluded us as a couple, so we elected to address our infertility problem by undergoing tests, and even some new procedures.

We eventually conceived—the natural way—but we always considered our daughters miracle babies even though technology was not part of the process (aren't all conceptions miracles anyway?). If necessary, would we have resorted to the still experimental procedures I used in the lab? We never had to answer that question, but many, many desperate couples do.

Now, more than fifteen years later, technology has continued to progress and the ethical questions are even more complex. Some reproductive options are so unusual that they border on the bizarre. Embryos can now be frozen and donated. There are women over the age of sixty conceiving babies from donor eggs; there are men who don't produce mature sperm becoming biological fathers; there are grandmothers who carry their own grandchildren in utero . . . the list goes on and on. And of course there's plain old IVF where Mom provides the eggs and Dad provides the sperm; the gametes meet in a dish, and the embryos end up back in Mom, who eventually gives birth to her own genetic child. This type of IVF produces more than 15,000 babies a year and it is considered quite routine.

How far should a couple be willing to go to have their own biological child? How much money should they spend? How much pain should they endure? (And it's not just the mother who may suffer in this type of parenting. Men of low fertility who opt for needle aspiration of sperm from the testis endure a painful process. I would guess that Lamaze breathing wouldn't help much in this procedure.)

On an almost daily basis we are exposed to reports in the media of amazing advances in biotechnology and biomedical sciences: the cloning of animals; sequencing of genes; new reproductive technologies; tests for defective genes; embryonic stem cell research; and development of genetically engineered food—to name just a few. We can react to these unprecedented developments on many different levels. From a practical perspective, we could ask: "how will these technological developments affect us?" From an emotional perspective we might ask: "am I ready to accept cloned people? Can I handle the knowledge about genetic defects in myself or my unborn baby?" From an ethical perspective we could wonder: "Is it ethical to know these things . . . or to perform such procedures?" Religious and secular ethical systems and codes of behavior have evolved to help people deal with all sorts of decisions.

Bioethical Issues Addressed by This Book

Brave New Judaism addresses how Judaism, an ancient religion, interprets and responds to recent advances in biotechnology and biomedical science. Compelling and unprecedented advances in biotechnology and biomedical sciences have presented society with baffling choices. What genetic diseases should we screen for in our unborn children? How far should an infertile couple go in their quest for a child? Should we be cloning humans? What is the status of a baby who was gestated outside of the womb? Should a woman have her breasts removed if she carries the breast cancer genes? Should we produce and eat bioengineered food? Should we use gene studies to define who is a Kohen (priest)? Jewish analysis of and reactions to these issues will help guide Jewish life in the future. The chapters that follow address some of the major issues in biotechnology and biomedical science.

Chapter 1 sets the stage for the discussion of cutting-edge issues by comparing and contrasting Jewish and secular bioethics. This chapter also introduces the theological and historical roots and development of halakha, or Jewish law. Eight common principles of Jewish bioethics are introduced and analyzed vis-à-vis Orthodox, Conservative and Reform thought. Applications of those principles appear throughout the book, with regard to various topics. (Readers who are familiar with some of the background information contained in chapter 1 may choose to skip some sections of this chapter and proceed to chapters 2 through 13, where main topics in biotechnology and biomedical science are explored.)

Reproductive issues are explored in chapter 2 (e.g., female infertility, donor eggs, disposition of eggs and embryos, establishing motherhood) and chapter 3 (e.g., male infertility, handling of sperm, testicular biopsy, artificial insemination and intra-cytoplasmic sperm injection). Chapter 4 raises questions about the production and use of human stem cells and the status of the early embryo in the Jewish tradition. The issue of human cloning is tackled in chapter 5: Is cloning permissible? What are family relationships of a cloned person? Does cloning fulfill the human obligation to procreate?

Genetic disorders found in Jewish populations are discussed in chapter 6 and genetic screening of embryos and fetuses is addressed. Chapter 7 presents the Jewish view of genetic engineering; artificial twinning, embryo freezing, and genetic alterations are explored. Ancient and high-tech methods of sex determination are discussed and secular and rabbinic reactions to sex selection are presented in chapter 8.

Chapter 9 reflects on the human genome project and its impact on determining Jewish genetic identity. The use of DNA testing to reveal Kohen genes, other Jewish genes, paternity, and the identity of human remains is explored. One issue addressed in chapter 10, "Judging Genes," is why Jewish populations appear to have so many flawed genes. This chapter also focuses on genetic screening of Jewish populations for breast cancer and other disease genes.

Questions regarding genetically modified (GM) plants and animals are tackled in chapters 11 and 12. Is the mixing of genes from different species permitted according to Jewish law? Can genetic engineering produce kosher animals from previously nonkosher species? Can genetic engineering render plants unkosher? Is environmental impact a factor in Jewish reactions to GM foods? Reflections on the genetic code are offered in chapter 13. Does Scripture allude to a DNA code? Are there analogies between Scripture and the DNA code?

On the one hand, many modern issues and advances appear to have no precedent in history or in religious texts. Indeed, even fifty years ago the most discerning scientists and theologians did not predict that by the turn of the millennium we would be able to decipher and alter genes, produce identical copies of animals at will, screen genes in order to predict the destiny of humans or produce new species of plants and animals for consumption.

On the other hand, although modern science is opening doors to novel phenomena that seemingly cannot be addressed by ancient texts, when one delves deeply into Scripture, sometimes we *can* connect religious perspectives with new findings. For instance, some have suggested that God "cloned" Eve from Adam's rib. The Matriarchs Sarah, Rebecca, and Rachel suffered from, and dealt with, their own infertility using prayer and the "technology" of their day (see chapter 2).

Science and religion are converging at breakneck speeds, hence this book is entitled: *Brave New Judaism: When Science and Scripture Collide.* Just as *Brave New World* by Aldous Huxley,[1] which envisioned a world of cloned humans, dealt with the specter of modern science intruding on humankind's basic right of reproduction, *Brave*

New Judaism addresses how modern biomedical science affects Jewish religious faith and principles. In fact, people of many faiths will need to address the unique questions posed by the convergence of cutting-edge science and their systems of beliefs, codes of conduct, and ethics.

Consult Your Rabbinic Authority

Jewish reaction to biotechnology is as diverse as the Jewish community itself. Thus, this book presents a wide array of information derived from many Jewish sources. It addresses Jewish views regarding complex issues from the perspective of biblical and talmudic sources, commentaries on these sources, and a variety rabbinic writings, decisions, and opinions. The talmudic sources cited here include halakha (Jewish law) and aggadah (a term used for parts of the Talmud that include stories, legends, and poetry—see chapter 1). It is important to keep in mind that, while aggadah is of interest and importance, it has a different status than halakha; aggadah is, traditionally, not considered binding as Jewish law.

Orthodox and some Conservative sources rely on the halakhic process. On the other hand, other Conservative sources and all Reform sources may make use of halakha to guide them, but not necessarily to govern them. In presenting a spectrum of sources and opinions, this book is not intended to fulfill the role of a rabbinic adjudicator or to serve as a source of halakhic rulings. I recommend, therefore, that any reader who requires a ruling in Jewish law should consult his or her own rabbinic authority.

Pushing the Envelope

How far should we go to keep a person viable? How far should we go to become parents? How far can we push the envelope and use the power of technology before we overstep our bounds and start to play God? Where do we draw the line?

As we learn in the ICUs, the ERs, and the ORs of our hospitals—even the most God-like specialists have limits to their medical miracles. Time after time we heard from the doctors that my mother's body must do the healing. Time and again she began to recover. And apparently only God—not doctors—can predict the time of death. When it was clear she was not going to recover, the doctors were unwilling to predict when she would pass on. Doctors do accept that they are not divine.

And when it's time for the beginning—a new life—to appear, this also happens in a mystical way. Doctors can facilitate the process, they can bring sperm and eggs together, they can even inject a sperm into an egg, but their power ends there. They can't explain or predict the whys and wherefores of conception; why conception works sometimes and not others. There are no medical guarantees in infertility research. Reproduction—the beginning of life—is apparently also in the hands of a higher force.

Many new high-tech developments are changing the way people ponder, plan, and live their lives. Childless couples pursue the route to high-tech parenthood. The ability to read and manipulate genes has led to novel opportunities to screen genes, select healthier offspring, correct defective traits, and eventually improve nature. Some worry that our attempts to improve on nature will lead to unforeseen problems. We are already getting a glimpse of this in the agricultural world. Biotechnologically engineered crops may be killing monarch butterflies and, even worse, some biotech corn meant for animal feed has contaminated human food supplies; there is insufficient research to determine whether this grain would pose health problems for humans.

The highly technological society we live in presents fascinating challenges for Jews. Historically, Scripture and other Jewish texts have documented Jewish reactions and responses to just about every conceivable scenario in everyday life, as well as extraordinary challenges such as wartime, illness, accidents, and death.

In the Talmud it is written, "Three keys are in the hand of the Holy One, blessed be He, and are not entrusted to the hand of any messenger: the key of childbirth, the key of revival of the dead and the key of rain." Thus, the rabbis were stating that birth, death, and weather were in the hands of God, and man will never be able to control those phenomena.[2]

The new technologies of the twentieth and twenty-first centuries have made possible what was formerly impossible, and we cannot predict the limits of these new powers. Although considered inconceivable in the recent past, we can now screen for genetic diseases, predict genetic destiny, and create identical copies of creatures in the laboratory. Other "impossible" scenarios that may soon be attained include: cloning humans, predicting with a high degree of certainty the time and cause of one's future demise, and controlling key elements of conception. These, and other tantalizing technological advances, will make life all the more complicated. In presenting modern advances from ancient perspectives, this book can help to elucidate the brave new Judaism that will emerge from these extraordinary times.

Acknowledgments

In writing this book, I have consulted with numerous sources ranging from the Torah, Talmud, and ancient commentaries to modern rabbis, scientists, and bioethicists. I have been privileged to have access to numerous erudite and thoughtful sources. I owe a huge debt of gratitude to the scholars whose works I consulted, and to the rabbis, scientists, and bioethicists who generously gave of their time.

A very special note of thanks to Phyllis Deutsch, editor at University Press of New England, whose insightful advice and comments helped me to formulate this book and to University Press of New England and Brandeis University Press for their support. I thank Mary Crittendon, Jessica Stevens, and Jennifer Thomas for their valuable contributions to the book's production.

I gratefully acknowledge the support of William Paterson University of New Jersey. I especially thank President Arnold Speert and Provost Chernoh Sesay. William Paterson University has been extremely helpful by providing sabbatical support and Assigned Research Time, which enabled me to apply the time, resources, and energy needed to complete this book.

I offer my deepest gratitude and love to my husband, Dr. Israel S. Wahrman, a psychologist who enjoys helping people. He is my editor extraordinaire, my *khavrutah* (learning partner), teacher, confidant, and best friend. I thank him for his constant support and encouragement and for more than a quarter century of love, friendship, counsel, companionship, partnership, lawnmowing, and landscaping.

I send my love and thanks to my terrific and talented daughters, Abby and Susie—for their patience, encouragement, inspiration, and support. My brilliant daughter Susie helped me to laugh when I needed it most. Her insightful questions helped me to see new connections. Abby's interest and pride in my work and her intelligent feedback gave me the confidence to keep plugging away.

My father, Rabbi Dr. Zev Zahavy, has served as my own personal rooting section since before I can even remember. I thank him for all that he has taught me about life, Judaism, and cosmology. His interest in and wonderment over the immense and mysterious universe inspired me to be curious about the natural world. His devotion to our faith provided a solid launchpad from which my own faith and love of Judaism was able to take off.

Edith Zahavy—my mother, of blessed memory—provided incessant encouragement and boundless love. She *qvelled* with pride and bolstered my confidence through every step of my career. She was always my best friend and my phone buddy, and the best role model a daughter could ask for. I do so wish that I could hand her this book.

My brother, Rabbi Dr. Tzvee Zahavy, encouraged and advised me when I first started writing articles on Judaism and science in 1997 for Jewish Communication Network. I thank him for all his brotherly and professional advice and for taking the time and trouble to critically read the draft of this book.

I thank my brother, Professor Reuvain Zahavy, for his special friendship and constant support, and for teaching me how to read when I was four. He helped me get through the roller coaster ride of ups and downs during the half year that our mother languished in the hospital. His brotherly and academic advice have helped me throughout my life. I also thank Bernice Zahavy and Dorothea Krieger for their valued friendship and moral support.

My dear mother-in-law, Sarah Wahrman, has inspired me with her courage. She documented her experiences during the Holocaust by publishing her account of survival during the most trying of times. My father-in-law, Rabbi Shlomo Wahrman, a prolific author, teacher, and scholar, continues to impress me with his erudite writings.

I thank Rabbi David Feldman, Dean and Founder of the Jewish Institute of Bioethics and Rabbi Emeritus of Teaneck Jewish Center, and geneticist Dr. David Weisbrot, Professor Emeritus at William Paterson University, for their critical reading of the manuscript and their helpful suggestions. My thanks also to genetic counselor Peggy Cottrell for her valuable input. I am grateful to Art Cottrell, who read the manuscript with a fine-tooth comb and helped me immeasurably to improve it. I thank both Peggy and Art for the many hours they devoted to their careful review of the book, as well as for their valued friendship.

My deepest gratitude to Rebecca Kaplan Boroson, editor of the *New Jersey Jewish Standard* and *Jewish Community News*, who encouraged me from the time she published my very first article on cloning. I am grateful for the support of James Janoff, publisher of the New Jersey Jewish Media Group, my colleagues Joanne Palmer, Helen Weiss Pincus, Alix Wall, and the entire staff of the NJJMG. I appreciate the helpful and supportive feedback from my readers in northern New Jersey, my electronic readers of America Online's (AOL) Jewish Community, and my colleagues at www.jewish.com. Note that some portions of this book have been based on articles I wrote for the New Jersey Jewish Media Group and other newspapers and magazines.

I thank my many colleagues at William Paterson University who have encouraged me and were supportive of my professional and research interests. I would especially like to thank Drs. Jaishri Menon, Gurdial Sharma, Jean Werth, and Eileen Gardner who collaborated with me on various research projects, and Drs. Peter Stein, Neil Kressel, and Melvin Edelstein, who encouraged my varied professional interests. I also thank Dr. Leonard Augenlicht, now of Montefiore Medical Center, who was my thesis advisor years ago at Cornell University. He taught me, through his example, to appreciate the discipline and the wonder of scientific inquiry.

And I thank Dolly the cloned sheep, and the scientists who created her. The existence of Dolly, the first mammal ever to be cloned, prompted and inspired me to begin writing on science and Judaism.

I take full responsibility for the contents of this book, and offer my apologies, in advance, for any and all errors, omissions, or misstatements.

June 2002   M.Z.W.

BRAVE NEW JUDAISM

CHAPTER I

Introduction:
Bioethics and the Jewish Spectrum

As a columnist for America Online, I have written about issues related to Judaism and science and, in turn, have received e-mail and even phone calls from Jewish and non-Jewish people from all over the world. Some of the questions sent to me concern a variety of diseases (Factor XI clotting deficiency, steroid dependent nephrotic syndrome, tyrosinemia, breast cancer, Hodgkin's disease, to name a few). I have heard from men and women experiencing various forms of infertility. In response to an article on honey, one Vermont woman wanted my opinion on the health benefits of maple syrup vs. honey. A Plano, Texas, man was interested in more information on Jewish astronauts. A woman from Australia asked me, "Do you think there is a link between art, science, and spirituality?" Some e-mails requested information on how to be tested for the "Priestly Gene." One woman wondered, "How do you clone a soul?" A few readers have even asked me how they could meet nice Jewish women or nice Jewish men.

The myriad of queries I have received in the past few years motivated me to write this book, which addresses numerous issues that provoke discussion and topics that will impact on the health and future well-being of humankind. Clearly there is a need to know about stem cells, cloning, genetic screening, infertility, bioengineering of plants and animals, and other cutting-edge technologies and how they relate to Jewish practice and thought. (This book does not, however, provide advice on how to meet prospective Jewish mates.)

In a climate of almost daily scientific breakthroughs it is vitally important to be equipped with tools to deal with the multitude of bioethical questions generated by these discoveries. Therefore, this book opens with a presentation of such tools, in the form of bioethical principles derived from traditional Judaic sources. Major bioethical principles are outlined in this chapter, and, throughout the book, these principles are cited with regard to relevant technologies and issues.

The second point revealed in this book is that, although some biotechnological breakthroughs have introduced problems and dilemmas that are truly unprecedented, the adoption and use of most forms of cutting-edge biological technology are frequently consistent with Judaic tradition and law. Finally, throughout this book I have analyzed Jewish reactions to biotechnology and biomedical advances from Orthodox, Conservative, and Reform perspectives. I demonstrate that different Jewish denominations have reacted to novel technologies in ways that are not always predictable.

Jewish vs. Secular Bioethics

Bioethics is "the critical examination of the moral dimensions of decision-making in health related contexts and in contexts involving the biological sciences."[1] Bioethical decisions can be based on religious principles, or on secular principles and theories, or on a combination of elements from both the secular and religious realms. In our secular society, most people would agree that you do not have to be a religious person in order to be an ethical person. However, "Even secular theoreticians find themselves looking to religious traditions—of which Judaism is by far the oldest—not always for definitive answers, but for insights that may inform their own decision making."[2]

Secular bioethicists make a point of distancing themselves from formal religions, or as described by Aaron Ridley, "sidestepping religion." This is an important precept for secular bioethics, since secular bioethics tries to encompass the views and practices of the whole society. Ridley explains, "If one person claims, for example, that abortion is wrong because the Pope says it is wrong, such a claim is clearly not going to carry weight with non-Catholics."[3]

Broad and general principles are developed for secular bioethics, so that they can become universal principles to guide people from a diversity of cultures and religions. "This doesn't mean that religious insights are unimportant for ethics. Quite the reverse," maintains Ridley. "But it does mean that we must find ways to express such insights in terms that are intelligible and persuasive to people whose religious perspective is different."[4]

Secular ethics provides for the development of codes of conduct for societies in which separation of church and state is considered important. In America, for instance, you don't need the Ten Commandments posted in every public school classroom in order to teach children not to steal and not to kill. It is already part and parcel of our legal system and of our social fabric. Youngsters are taught in kindergarten to respect the feelings and property of others. School children are taught in secular environments, by their parents, teachers, and other role models not to harm others, and to treat people kindly.

By keeping religion out of the discussion of bioethics, it is possible to approach bioethical dilemmas from a common ground to which everyone can relate. Ridley declares, "A powerful motivation for engaging in philosophical argument about ethical matters, therefore, is the desire to avoid the deadlock and futility characteristic of religious disagreement."[5]

Although scholars of secular bioethics emphasize that sidestepping religion is important for dealing with ethical problems in a diverse society, in fact, many of the major secular principles are, indeed, anchored in or derived from religious codes. While modern secular bioethics and religious ethics are considered distinct from one another, they share many of the same principles and doctrines.

Principles and Case Studies

Principle-based approaches to ethics are top-down ("here are the rules, follow them"), while casuistry is a bottom-up type of ethical approach that involves the application of case studies ("here are the situations, figure out the rules for yourself"). When either approach is used exclusively, there is limited flexibility and adaptability to new situations. Thus, combination approaches are more acceptable to many ethicists. One example of a combination approach is called Reflective Equilibrium. Developed by John Rawls, this approach combines theories, principles, rules, and judgments about specific cases.[6] Many legal and ethical systems are based at least in part on casuistry. For instance, legal precedents play important roles in determining decisions of American courts. Rabbinic Judaism relies heavily on case-based reasoning as well; rabbis typically cite past decisions in order to determine acceptable norms of behavior. But legal and ethical systems also combine case-based reasoning with a clear set of rules. The Judaic system of halakha (Jewish law) combines doctrines with case-based reasoning.

Rabbi Elliot Dorff indicates the importance of balance between general rules and individual cases to achieve the flexibility needed in real life. Dorff states, "Making decisions in specific cases does not eliminate the importance of articulating general standards—that is, commonly used principles and policies; one must just know when and how to use them."[7] For example, Dorff offers, "Jewish law establishes a clear general *principle* to preserve life; but like all other principles and policies, this one is open to being supplanted in given circumstances by specific considerations."[8] For instance, the principle "thou shalt not kill" may be supplanted in times of war with permission to kill one's enemies as a defensive act.

Judeo-Christian values serve as a foundation for many secular ethical principles, and thus, the Jewish view on bioethics coincides in many instances with secular principles. However, although Rawls's Reflective Equilibrium has features that are analogous to the halakhic model in its structure, the former is secular and the latter is based on religious principles. Even within Judaism there are significant differences of opinion. As we will see in analyzing modern biotechnological issues, Reform, Conservative, and Orthodox rabbis do hold different opinions with regard to a variety of bioethical dilemmas.

The religious spectrum, consisting of the main denominations that exist in Judaism today, include (going from "right," most traditional, to "left," most liberal): Orthodox, Conservative, Reform, and Reconstructionist. The methodologies used to approach bioethical issues vary in the different denominations, just as their approaches to Judaism differ. However, vis-à-vis bioethics, it is interesting to note that, while there is not universal agreement on all issues, there are many consistent rabbinic viewpoints and decisions among the various denominations.

A universal principle found throughout the main denominations involves Jewish law being rooted in Divine Authority. "Jewish law ultimately gains its knowledge and authority from God," writes Michael Grodin, a professor and bioethicist at the Program in Bioethics and the Institute of Jewish Law at Boston University. "As opposed to secular law, which is primarily grounded in autonomy and civil rights, Jewish law is based on religious metaphysics and epistemology. Halakha focuses on a wide range of covenental obligations, roles, and responsibilities. A shared sense of the common good forms individual and communal bonds, as one develops relationships with other members of society and with God. For 3,000 years Halakha has been applied to all aspects of Jewish life, including questions of civil, criminal and religious nature."[9]

Development of Halakha

The major texts on which halakha is based include the Torah and its commentaries, the Talmud (including the Mishnah and Gemara), dozens of commentaries on these tomes, and layer upon layer of rabbinic response to questions arising in the Talmud and questions arising from day-to-day situations and events. The evolution of halakha spans three thousand five hundred years, from Moses and his disciples through the generations of men who are recognized as major recorders, codifiers, and commentators: the *Tannaim*, the *Amoraim*, the *Geonim*, the *Rishonim*, and the *Aharonim*. Approximately 140 generations have passed since Moses ascended Mount Sinai. According to tradition, the line of transmission has not been broken, and the enormous literature has grown and grown to the point that no one person can master all that has been written on the Torah in his or her lifetime.

The compendium of material on which modern bioethical and other Jewish legal decisions are made includes the Written Law and the Oral Law. The Written Law includes the Torah, also known as the Five Books of Moses or the Pentateuch. The Oral Law includes a corpus of doctrine, correlative to the Written Law, traditionally believed to have been communicated from Moses through a chain of successors to the authors of the Mishnah. In the Mishnah it is written: "Moses received the Law from Sinai and handed it down to Joshua, and Joshua to the elders, and the elders to the prophets, and the prophets handed it down to the men of the Great Assembly."[10] The "elders" are understood to be the judges *(shoftim),* including some of the colorful and dynamic leaders who succeeded Joshua: Othniel, Deborah, Ehud, Gideon, and Samson. Among the dozens of prophets who followed them were the spirited and tenacious Samuel, Elijah, Isaiah, Jeremiah, and Ezekiel. Toward the end of the period dominated by the prophets, the great teacher and leader Ezra, who together with Nehemiah led the Jews from Babylonian exile back to Israel, established the Great Assembly. This group consisted of 120 men, including teachers, scholars, and prophets, who helped to maintain the continuity of the transmission and interpretation of the Oral Law.

The *Tannaim* (teachers) were a group of rabbis whose scholarship led to the codification of the Oral Law in the Mishnah. Two rabbis in particular, the esteemed scholars Hillel and Shammai, heads of two separate schools of *Tannaim,* greatly influenced the writings of the Mishnah. Generally the School of Hillel adopted a more lenient view and the School of Shammai developed more stringent interpretations and rulings, and in a great majority of cases, the normative halakha was decided in accordance with the view of the School of Hillel.[11]

Yehuda HaNasi, or Judah the Prince, who was born in 135 C.E., is credited with compilation of the corpus of Jewish law into the Mishnah. The written form taken by the Mishnah included a record of the discussions and opposing views of the rabbis. The Mishnah is divided into six sections, called *Sedarim* (Orders); each Order is made up of several subdivisions, called *Massechtot* (Tractates). There are a total of sixty-three tractates. The Orders are: *Zeraim* (Seeds), *Moed* (Seasons), *Nashim* (Women), *Nezikin* (Torts), *Kodashim* (Sanctities), and *Teharoth* (Purities). In addition, there are a number of tractates of post-Mishnaic origin that are not included in the six Orders. Mishnaic Tractates cover far-ranging topics including Sabbath and other festivals, issues of marriage and divorce, property and inheritance, Temple laws and customs, and a myriad of other subject matters. The style of the Mishnah helped to stimulate further discussion and study, which eventually resulted in the writing of the Gemara.[12]

During the next few hundred years scholars, called *Amoraim* (speakers), studied and debated the Mishnah and developed detailed commentaries on it. The *Amoraim* studied in academies based in Babylon and in Israel, producing two separate works. Although the two groups worked mostly independently of each other, there was some exchange of ideas via rabbis who traveled between the two communities. The end result of this scholarship was the Gemara, which includes the Babylonian Talmud and the Jerusalem (or Palestinian) Talmud. The Babylonian Talmud goes into greater depth than its Jerusalem counterpart on many issues and is more than seven times the length of the Jerusalem Talmud. Accordingly, it is the Babylonian Talmud that most students and scholars study. (Note that most of the talmudic citations recorded in this book are from the Babylonian Talmud.)

The Talmud is made up of two types of text: halakha and aggadah. Halakha is from the Hebrew word *halokh* (to go or to walk); it defines a code of action or a way of life. Aggadah, on the other hand, includes stories and poetry, satire, metaphor, word and number symbolism, and moral tales. Some aggadic literature was incorporated into the branch of rabbinic literature called *midrash*. The great corpus of halakha and aggadah found in the Talmud has been studied and analyzed over the past 1500 years resulting in layer upon layer of commentary.

From the sixth to the eleventh centuries, during the age of the *Geonim* (the Hebrew term for eminences), prominent rabbinic leaders produced extensive post-talmudic responsa. Shortly after that period, Rabbi Shlomo Yitzhaki, best known by the Hebrew acronym Rashi, rose to become one of the greatest biblical and talmudic commentators of all times. Residing in France from 1040 to 1105, he was considered one of the first *Rishonim* (early commentators).[13] His commentary on the Bible became popular all over the world as it addressed complex passages in the text and clarified important issues. Today in modern yeshivas even young children study Rashi's commentary on the Bible. By the fourth or fifth grade children learn how to decipher the semi-cursive "Rashi script" (used for printing other commentaries as well). They learn to analyze difficult passages in the Bible by asking "what's bothering Rashi?" and they discover his lucid and insightful answers, which illuminate and explain complexities in the Torah. Rashi's commentary on Talmud is likewise invaluable in terms of its insights, explanations of obscure passages, and revelations on the intricacies of talmudic discourse.

Rashi's grandson, Rabbi Yaakov ben Meir (known as Rabbenu Tam), continued to dissect talmudic arguments and produced commentaries that are studied and revered today. Writings of Rabbenu Tam and other commentators make up the *Tosafot* (Supplements).

Rabbi Moses ben Maimon, or Maimonides (also known by the Hebrew acronym Rambam), who lived in the eleventh century, produced the *Mishnah Torah,* a gigantic compendium of Jewish law arranged in fourteen parts.[14] Other important works of Maimonides include his *Guide to the Perplexed* and his *Commentary on the Mishnah.* Since Maimonides was not only a brilliant commentator, but also a renowned physician, his insight into bioethical issues serves to elucidate numerous issues relevant to modern biomedical science. Many of his insights have relevance to discussions on cutting-edge Jewish bioethics.

Rabbi Moses ben Nahman—also called Nachmanides (or his Hebrew acronym Ramban)—was a preeminent scholar of the thirteenth century who wrote commentaries on the Bible and on Talmud. Later commentators of importance also include Rabbi Menahem HaMeiri whose insightful commentary covers most of the tractates of the Gemara.

Rabbi Joseph Caro, a sixteenth-century codifier of Jewish law, who was expelled from Spain with his parents in 1492 and ended up in Safed, Israel, compiled the classic code of Jewish law, the *Shulhan Arukh* (the "set table") in 1565. This work contained the collected views of previous Sephardic scholars and included the four categories: *Orah hayyim* (way of life)—Sabbath, festivals, daily prayers; *Yoreh Deah*—dietary laws, vows, purity; *Even HaEzer*—marriage, divorce, sexual relations; *Hoshen HaMishpat*—civil and criminal law, inheritance, property. Rabbi Moses Isserles of Poland (Hebrew acronym, Rama) subsequently added his compilation, *Ha-Mappah* ("the tablecloth"), to complete the work, by including halakhic decisions of Ashkenazic scholars.[15] Later commentaries on the *Shulhan Arukh* and *Ha-Mappah* were written by the *Aharonim* (the last ones). The *Kitzur* [abridged] *Shulhan Arukh* (1863), which was written by Rabbi Solomon Ganzfried, was meant to make the laws of the *Shulhan Arukh* more accessible to laymen. The *Arokh HaShulhan,* written by Rabbi Yechiel Epstein a few years later, provides insight into Jewish law from a late-nineteenth-century perspective.[16]

Rabbinic commentators of today frequently consider their own works as arising from "standing on the shoulders of giants." It is traditionally held that the older the work, and the closer it is to Moses and Sinai, the more stature it has. Thus the *Tannaim* are considered giants, dwarfing the *Amoraim. Amoraim* overshadow the *Rishonim,* and *Rishonim* outweigh the *Aharonim* in terms of intellectual stature and insight. (A biological analogy could be a mite on the back of a flea, on the back of a dog . . . etc.) The layers upon layers of commentaries have evolved into an intricate, interrelated body of legal, ethical, mystical, and practical analysis of the Torah.

Halakha and Bioethics

Today, halakha represents a legal system based on Scripture, which serves as a guide for living in a modern world. Thus, halakha—at once ancient, but also constantly evolving, is able to address complex and sometimes novel issues.

Michael Grodin compares Jewish law to secular legal systems. "Jewish law has the equivalents of a constitution, codes, statutes and case law. Contemporary Jewish law has legal authorities, courts, and an enormous, indexed legal literature." The "constitution" is, of course, the Torah. "Just as it would be virtually impossible to base current secular legal decisions solely on the text of the U.S. Constitution, however, it has been necessary for the Torah to be supplemented with an extensive array of legal interpretations," states Grodin. The Talmud represents the major source for interpretation and case analysis; responsa, written by rabbis throughout the ages, represent further rabbinic interpretation of a myriad of cases. Grodin proposes that responsa are "somewhat analogous to 'common law' decisions in Anglo-American secular law." [17]

The process of development of halakha has involved controversy and debate. Throughout the Talmud conflicting views of scholars are presented. Grodin notes "that disagreement and argument are not unusual in Jewish law. In fact, they are the norm. In Halakha there is not one central authority that ultimately establishes a binding universal answer to a specific question. Precedents become accepted through interpretation and deference to great Torah scholarship. Custom frequently acquires normative weight." [18]

Branches of the Judaic Tree

So it is not surprising that, when it comes to bioethics, there are conflicting views on many issues. At a recent conference on Torah and science, Dr. Louis Flancbaum observed that all branches of Judaism are valiantly trying to come to terms with new advances in biotechnology and biomedical sciences. "All branches of Judaism have Responsa," he remarked. He continued that there is a "tremendous amount of thought which goes into these Responsa." He suggested that, "In the ethical arena, these Responsa show that the differences are more perceived than real. Almost without exception, the Reform and Conservative Responsa on bioethical subjects parallel moderate Orthodox opinions, while within Orthodoxy, there often exist opinions that occupy opposite ends of the halakhic spectrum."[19]

Flancbaum further suggests that while Conservative and Reform are more centralized and the centralized authorities make decisions that are more consensus-based, since Orthodoxy is decentralized, Orthodox opinions appear to be more diverse. In other words, even though Orthodoxy is considered to be more bound by ancient precedent, surprisingly, "in areas of bioethics often Orthodox opinions give opposite stances."[20] This is seen over and over again in the bioethical issues discussed in this book. The wide range of opinions from the Orthodox camp are sometimes astounding, considering that they are based on uniform and accepted texts. It is also interesting to note that, in issues of bioethics, frequently Orthodox authorities seem more accepting of modern technologies than their Conservative and Reform counterparts.

Grodin's article addresses the different approaches to bioethics that are characteristic of the major denominations. He explains that "in addition to the Orthodox approach to Jewish medical ethics . . . there are other approaches. In the U.S., 90 percent of Jews are part of either the Conservative, Reform or Reconstructionist denominations or are unaffiliated. Despite their majority in numbers, these branches of Judaism have a much smaller collection of writings or rabbinical *Responsa* on medical ethics [than the Orthodox]."[21] Orthodox rabbis have been particularly prolific on biomedical issues and have analyzed many scenarios, including those novel areas explored in this book. Only a handful of Conservative and Reform rabbis have addressed these questions in a formal way. However, as noted above, the centralized authorities of the Reform and Conservative movements have established review boards to determine policy based on rabbinic analysis.

Rabbi Elliot Dorff explains the process adopted by Conservative rabbinic authorities. "Authors affiliated with the Conservative move-

ment have been much more aggressive than their Orthodox colleagues in asserting a major role for ethical concerns," explains Dorff. "But unlike their Reform counterparts, Conservative rabbis use moral values as an integral part of the Jewish *legal* process by which contemporary decisions should be made."[22]

On the other hand, according to Grodin, the Reform approach incorporates halakha, but then allows for further non-halakhic analysis. "The Reform or liberal movement of Judaism also allows for halakhic approach to Jewish law. Reform Judaism, however, does not see itself as bound by Halakha and allows for an extra-halakhic Jewish ethical analysis, as well. Thus, while Rabbi Solomon Freehof has written extensive Reform responsa on questions of bioethics, individual Reform Jews would be able to interpret and analyze for themselves the materials as presented within the tradition."[23] One responsum of Reform Judaism's Central Conference of American Rabbis (CCAR) reaffirms Grodin's analysis of their viewpoint. "Rabbinic scholars ought to acknowledge that traditional techniques of halakhic analysis, in particular the case method of reasoning by analogy, are of limited usefulness in an area dominated by technological novelty and innovation. The tortuous logic of the arguments we have just cited demonstrates that there may simply be no precedents or source materials in Talmudic literature that offer plausible guidance to us in making decisions about these contemporary scientific and medical issues."[24]

Grodin describes the Reconstructionist approach to bioethics by explaining that "the Reconstructionist movement of Judaism, which views Judaism as a civilization, has less developed literature on Jewish bioethics. There are also Judaic approaches to questions of medical ethics that focus less on the concept of precedent in Jewish law and rely on a covenental approach."[25] (There are few written sources on Reconstructionist views of bioethical issues; accordingly, this book will not address the perspectives of that movement.)

Main Principles of Secular Bioethics

Some aspects of secular ethical systems have features in common with Jewish ethical thinking and processes. Many major ethical principles are rooted in biblical commandments. Principles of secular bioethics also share commonalities with Jewish bioethical principles—and may

even be rooted to some extent in Judaic thought. Thus, analysis of bi-oethical dilemmas from a Jewish standpoint would appear to be relevant to Jews and to many others within the secular world as well. Secular principles of bioethics, which have been developed specifically to address issues in medical and environmental science, are listed below. It is interesting to note analogies and possible derivations of these ideas from Judaic sources.

Four main principles of modern secular bioethics have been identified and have served as cornerstones for development of bioethical codes of behavior. Beauchamp and Childress outlined these four principles. They are: respect for autonomy, nonmaleficence, beneficence, and justice.[26]

1. *Principle of respect for autonomy.*

This includes freedom of choice and liberty of action. From this principle, the right of informed consent was developed. The patient's autonomy is of paramount importance; patients should be educated and allowed to participate in decisions regarding their fate. Patients should retain authority to determine what their course of treatment is.

Although not synonymous with autonomy, human free will has many features in common with that bioethical principle. And the exercise of free will is a major principle in Judaism. Human choice and culpability are discussed in Deuteronomy 30:19: "I have placed life and death before you, blessing and curse; and you shall choose life so that you will live, you and your offspring." Rabbi Hanina Ben Hama pronounces: "everything is in the hand of heaven except the fear of heaven."[27] God may have omniscience, but "God's providence does not apply to human moral deliberations."[28] Rabbi Akiva states in Mishnah *Avot*: "Everything is foreseen by God, yet man has the capacity to choose freely."[29] Since free will is such an important concept, it stands to reason that patients should be afforded input regarding their fate. However, even patient autonomy has limits; for instance, many would agree that patients must be prevented from harming themselves. Euthanasia on demand is not legal in the United States.

2. *Principle of nonmaleficence.*

Do no harm; do not cause injury to patients.

The Ten Commandments declare "thou shalt not kill". In addition, the Talmud states, "Therefore only a single human being was created

in the world, to teach that if any person has caused a single soul of Israel to perish, Scripture regards him as if he had caused an entire world to perish; and if any human being saves a single soul of Israel, Scripture regards him as if he had saved an entire world."[30] There is clear injunction in Jewish law to refrain from harming others. Later in this chapter more examples are offered on this point in the section on *pikuakh nefesh*, the preservation of life.

3. Principle of beneficence.

Do good; promote patient welfare.

The obligation to save a life of someone in danger is derived from the passage "Nor shall you stand idly by the blood of your fellow."[31] "And heal, he shall heal" (Exod. 21:19–20) provides for the obligation to practice medicine. (The Talmud says "From here [it is derived] that the physician is granted permission to cure."[32]) "And you shall restore it to him" (Deut. 22:2) refers to restoring the health of a fellow man as well as his property. More details on this are presented later in this chapter, in sections on *pikuakh nefesh* (preservation of life) and *v'rappo yirappe* (and heal, he shall heal).

4. Principle of justice.

Social benefits (such as health care services) and social burdens (such as taxes) should be distributed in a just manner.

"Tzedek, tzedek tirdof," "Justice, justice shalt thou seek" (Deut.16: 20), is an explicit scriptural commandment to pursue justice. "You shall not commit a perversion of justice" (Lev. 19:15) refers to the equality of all persons before the courts of justice, without any show of favor to either the poor or the mighty (also see Deut. 1:17). Human justice is modeled on divine justice (*imitatio Dei*). Scripture explains that all people are not really equal; some should be favored, namely the unfortunate victims of our society: widows, orphans, strangers, the poor, and the needy. Those individuals are singled out for special treatment as they can easily be exploited and used by unscrupulous individuals.[33]

As seen above, in many instances Jewish principles support and help to define secular principles. A major difference between Jewish and secular principles, however, is revealed by a statement of Rabbi Elliot Dorff: "The classical rabbis made it very clear that the Jewish tradition they were shaping . . . rests . . . in the hands of the rabbis of each generation who interpret and apply it according to their own understanding of God's will. . . . This fundamental commitment of Jewish law to morality, based on an underlying theology of a moral God who commands it, should also guide us as we face the many new issues in contemporary bioethics."[34]

Fringes of the Same Tallit: Judaic Bioethics

From the perspective of one fringe of the *tallit* (the Jewish prayer shawl)—the most observant, traditional Jew—to the opposite fringe of the *tallit*—the most liberal, nontraditional Jew, Jewish reactions to recent advances in biotechnology and biomedical science have many commonalities. Despite the diversity in the Jewish world today and varied approaches to Jewish life, many Jews share a common theology; common principles underlie Jewish thought.

In discussing Jewish denominational similarities and differences regarding biomedical science, Michael Grodin states: "Several normative principles of Jewish law that relate to medicine have evolved over the centuries. These principles include the right and duty to treat patients, the duty to seek care, and the sanctity of human life."[35] Major principles on which Jewish bioethics are based can be traced first to scriptural passages and ideals. I present below eight major cornerstones of Judaic bioethics, which help to guide the interpretation and evaluation of new scientific discoveries. These represent common principles adhered to and supported, to varying extents, by different denominations of Judaism. The sources for these principles are found in the Written and Oral Torah, and that is why they may be identified as distinctly Judaic principles of bioethics.

The common principles are:

1. Be fruitful and multiply; fill the Earth and subdue it—*Pru ur'vu, umilu et haaretz v'kivshuha*
2. Concern for the suffering of animals—*Tza'ar baalei khayim*
3. Repairing the world—*Tikkun olam*
4. Do not destroy—*Ba'al tash-khit*
5. Preservation of life—*Pikuakh nefesh*
6. And heal, he shall heal—*V'rappo yirappe*
7. According to the natural way—*Derekh hateva*
8. There is nothing new under the sun—*Ayn khadash takhat hashemesh*

1. Be fruitful and multiply; fill the Earth and subdue it—*Pru ur'vu, umilu et haaretz v'kivshuha*

The Bible directs humankind: *"P'ru ur'vu u'milu et haaretz v'kivshuha,"* (Fill the Earth and subdue it . . . have dominion over every living thing . . .) (Gen. 1:28). This is the first commandment in the Bible, and Jewish authorities from across the spectrum view this passage as a "Prime Directive." To a varying extent Jewish leaders support and encourage establishment of the Jewish family. Orthodox sources with few exceptions prohibit most forms of birth control and emphasize the requirement to have at least two children. More liberal interpretations are less rigorous in interpreting this as a requirement, but use this passage to support the centrality of family life in the Jewish community.

For the purposes of Jewish bioethics, this passage appears to give humankind free rein to utilize, and even exploit other living things on our planet. An important question to pose here is: What is meant by dominion? Our power to apply new technologies needs to be defined more clearly so we can understand the limits of human authority over Earth and nature.

Are there limits to this power? Is it permissible to harness technology to create better babies through genetic engineering? Should correction of genetic disease be in our future? Should we use the technology for cosmetic, or perhaps even frivolous reasons? Should we make our children taller, more athletic, more intelligent? Should we produce and eat genetically engineered plants and animals? The powers to change living creatures are within our reach, but which applications are necessary? Which are permitted? Which are unethical? Which are abhorrent?

"Be fruitful and multiply," which is seen as the first commandment or mitzvah presented to humankind, is taken quite literally by Orthodox interpretations to indicate the obligation of Jewish males to father children. This passage has an impact on many applications of biomedical science, since it affects decisions made in human reproductive technology (see chapters 2 and 3). "And subdue it" indicates humankind's domination of other earthly organisms and the environment. The boundaries of this power are defined in later biblical passages that provide restrictions on human dominion. One guideline, for instance, stipulates that the act of dominion must not violate any other halakhic injunctions.

2. Concern for the suffering of animals—*Tza'ar baalei khayim*

Dominion over animals is an issue of importance when animal cloning and other use of animal tissue are considered. Walter Jacobs states in the Union of American Hebrew Congregations (UAHC) Program Guide on "Cloning": "The Biblical statement in Genesis (1:26) placed people above animals and enabled them to rule them and therefore to use them in any way that seemed appropriate and certainly to save a life (*pikuakh nefesh*)."[36]

Jacobs gives examples of how animals may be used: cattle may be kept for food or work, animals may be sacrificed as part of Temple worship, and wild animals that pose a danger to the community may be killed. However, boundaries are clearly set, emphasizing humane treatment of animals. There are directives on how to catch and house animals, prohibitions against eating a limb from a living animal, and prohibitions against killing a mother animal with its young.[37] Jacobs describes a list of stipulations on use of animals: "Kindness to animals included the lightening of the load from an over burdened animal. . . . [D]omestic animals were required to rest on *shabbat* as human beings. . . . If an animal needed to be rescued it was to be done even on *shabbat*. . . . [T]he castration of animals was prohibited and this has always been considered as a form of maiming, which was forbidden. . . . They may be used by human beings but not treated cruelly."[38]

Jewish law permits the use of animals in medical research, because the noble goal of *pikuakh nefesh*, preservation of human life, supersedes the issue of animal suffering. However, halakha insists that any suffering of animals be kept to a minimum. Animals may be slaughtered for food, but the knife used should be as sharp as possible so that the animal suffers the least pain possible.[39]

Clearly there is a dynamic tension between the first commandment found in Genesis—giving humankind dominion over every living thing—and our duty to act humanely with regard to other creatures. Thus, rabbinic scholars of all denominations have grappled with the question of boundaries and have tempered the application of human power over creatures and the environment with the principles of *tza'ar baalei khayim*, *tikkun olam*, *ba'al tash-khit*, and *derekh hatevah*.

3. Repairing the world—*Tikkun olam*

The "Statement of Principles for Reform Judaism" indicates: "Partners with God in *tikkun olam,* repairing the world, we are called to help bring nearer the messianic age. . . . We are obligated to pursue *tzedek,* justice and righteousness . . . to protect the earth's biodiversity and natural resources. . . . We reaffirm social action and social justice as a central prophetic focus of traditional Reform Jewish belief and practice."[40] *Tikkun olam* has taken on a renewed importance, strongly emphasized in Conservative and Reform principles.

Orthodox sources also discuss the act of "perfecting the world" as derived from Nachmanides (on Gen. 1:28). "And conquer it—God gave man power and control on earth to do as he wishes with the animals and insects and everything which crawls on the earth, and to build, to uproot what is planted, to quarry copper from the mountains, etc." However, the stipulations are that any act must not violate a halakhic prohibition, and the "act of improvement must bring benefit to human beings, or at least a measure of benefit which exceeds the damage caused."[41]

In many of the issues discussed in this book, *tikkun olam* is a critical question. Some might argue that using our power to "fix" genes or produce new species is an example of repair of the Earth. Others would maintain the exact opposite position: that biotechnology is exploitative of the natural world; we are not "fixing" anything, we are meddling in areas not meant to be changed. It all depends on how the powers of biotechnology are considered: Are they powers of nature just waiting for humans to discover and use? Or are they a perversion of nature? One could argue that biotechnology is inherently neutral, and it is up to humans to decide how to apply these powerful tools.

4. Do not destroy—*Ba'al tash-khit*

Ba'al tash-khit was derived from Deuteronomy 20:19, a passage that speaks against destruction of fruit-bearing trees and that was extended by the Rabbis to forbid the wasting of all useful things. The great scholar and physician Maimonides wrote: "It is not only forbidden to destroy fruit-bearing trees but whoever breaks vessels, tears clothes, demolishes a building, stops up a fountain or wastes food, in a destructive way, offends against the law of 'thou shalt not destroy.'"[42]

Other examples of *ba'al tash-khit* include hunting for sport, which was strongly condemned. Indeed, only two biblical figures, Nimrod and Esau, are identified as hunters; and this vocation was not lauded—it was considered anti-Jewish. Likewise, it is not permissible to take a bird from her nest along with her young. It is forbidden to slaughter a cow and her calf on the same day. Nachmanides wrote (regarding Deut. 22:6), "he who kills the mother and offspring on one day is considered as if he destroyed the species."

The concept of *Ba'al tash-khit* should help to guide the applications of many biotechnological advances. For instance, it might be permissible to produce an oil-digesting microorganism to be used to clean up an oil spill. Could the same organism be used in warfare to destroy the enemy's fuel supply? Or would that be considered destruction of something of value?

5. Preservation of life—*Pikuakh nefesh*

One of the fundamental beliefs guiding Jewish medical ethics is *pikuakh nefesh*, preservation of life. Talmud *Sanhedrin* discusses cases in which one is obligated to save a life, including reviving a person from drowning.[43] From these principles we derive the obligation to practice medicine, to use natural resources, to develop new technologies, and to go so far as to violate the Sabbath, kashrut (Jewish dietary laws), and other cornerstones of religious observance in order to preserve life.

Maimonides wrote: "It is commanded that we violate the Sabbath for anyone dangerously ill. One who is zealous [and eagerly violates the Sabbath in such a case] is praiseworthy; one who [delays in order to] ask [questions about the Law] is guilty of shedding blood." The Talmud states this principle in a number of instances: "Preservation of life overrides all other considerations"; and "Sakanat nafshot docheh et hakol" (danger to life overrides all other obligations). It is obligatory to preserve life at all expense, and that includes a person's own life. Personal martyrdom is forbidden except when the alternative is idol worship, taking another person's life, or performing adulterous sexual acts.[44]

Rabbi Elliot Dorff, writing from a Conservative Jewish perspective, states, "Jews have the duty to preserve their own lives. When interpreting Leviticus 18:5, which says that we should obey God's commandments 'and live by them,' the rabbis deduce that this means that we should not die as a result of observing them. . . . If, however, Jews need to violate Sabbath laws or steal something to save their own lives or the life of someone else, then they are not only permitted but *commanded* to violate the laws in question to save that human life. . . . [S]aving a life is the most sacred of obligations."[45]

Jacobs also emphasizes that "Pikuach nefesh is an overriding consideration. . . . Human life must be saved if it is at all possible and even some pain to animals is permitted for this purpose."[46] In addition, the Reform Jewish Union of American Hebrew Congregation's "Program Guide IX: Organ Donation and Transplantation" states unequivocally, "The inestimable value of human life is a cardinal principle of Jewish law." Rabbi Richard Address and Dr. Harvey Gordon write in their introduction to the document, "We are fortunate to live in a time when technology can make it possible to act on the concept of *p'kuach nefesh* in ways hardly dreamed of just a few years ago."[47] In 1996 the Women of Reform Judaism developed a program called "*Matan Chaim*: The Gift of Life," which provides an organ donation brochure and cards, along with educational program guides to help launch organ donation programs within congregations. This is just one example of the use of modern medical science to support the principle of *pikuakh nefesh*.

The Orthodox view has been reiterated clearly by many writers. Rabbi J. David Bleich sums up the Orthodox position: "Human life is not a good to be preserved as a condition of other values but an absolute, basic and precious good in its own right. The obligation to preserve life is commensurately all encompassing."[48]

The Torah's commandment "you shall not stand idly by the blood of your neighbor" (Lev. 19:16) is interpreted by rabbis as the obligation to save others. Other passages exhort Jews to guard their own safety and lives. "Be careful and look after yourself scrupulously" (Deut. 4:9) and "Look after yourself and take care of your life" (Deut. 4:15) have been used by rabbis to forbid smoking and other potentially harmful behaviors.

6. And heal, he shall heal—*V'rappo yirappe*

"And heal, he shall heal" (Exod. 21:19) implies that medical advances may be used therapeutically to improve the treatment of patients.[49] How have Judaic scholars reacted to the breakneck developments in biotechnology and biomedical science? This passage is crucial in determining the permissible applications of many new developments in health and medicine, including experimental and nontraditional approaches to disease.

The talmudic proclamation "If any human being saves a single soul of Israel, Scripture regards him as if he had saved an entire world"[50] has encouraged Jews throughout the ages to practice healing arts. There have been lively rabbinic debates dating back many centuries, in which some rabbis challenge the right of humans to heal the sick, in essence reversing the natural order. They argue that disease and infirmity are also determined by God—and who are we to challenge God's will? The prevailing arguments, however, maintain that the powers to heal, the powers to harness science to improve the human lot, are also God-given, and as such we are permitted, and probably required, to use these God-given tools to improve the world.

According to Maimonides, "And you shall restore it to him" (Deut. 22:2) also refers to the obligation to restore a man's health (as well as his possessions).[51] And Nachmanides evens finds the obligation to heal inherent in the Golden Rule—"And you shall love your neighbor as yourself" (Lev. 19:18).

7. According to the natural way—*Derekh hateva*

What is considered natural, and what is unnatural? *Kilayim*, or the mixing of species, is unnatural, according to Nachmanides, who asserts that the grafting of different species results in inappropriate mixing of species created by God, in essence changing the order of creation. This may become an important point where manipulation of species using new technologies is considered (see chapters 11 and 12).

Kishuf, magic and sorcery, are clearly taboo, as they run counter to the principle of *Derekh hateva*. Exodus 22:17 states, "You shall not permit a sorceress to live." Deuteronomy 18:10–11 commands, "There shall not be found among you one who causes his son or daughter to pass through the fire, one who practices divinations, an astrologer, one who reads omens, a sorcerer, or an animal charmer; one who inquires of Ov or Yidoni, or one who consults the dead."

Rabbi Immanuel Jakobovits, often called the father of modern Jewish bioethics, writes on the role of the supernatural in Judaic thought:

"by and large, Jewish law, where it did not altogether proscribe super-stitious practices, at best tolerated them as a concession to human ad-diction. It found very little space for the faith-healer, and none at all for the professional quack—the favorite character in the medical legis-lation of the past millennium and more. It knew nothing of healing shrines or relics, and next to nothing of the exorcism of demons."[52]

Despite the strong taboos against the supernatural, there are occa-sions where magic may be acceptable, since Judaism cherishes life above all else. As explained by Jakobovits, "Jewish law treasured the protection of human life so intensely that it was prepared, as a general rule, to give the accepted claims of magic and the occult virtues, how-ever questionable, the benefit of the doubt, often even at the expense of its own religious injunctions. For whenever law and life are in con-flict, Judaism usually shows a strong bias in favour of life."[53]

8. There is nothing new under the sun—*Ayn khadash takhat hashemesh*

According to Ecclesiastes (1:9), everything we see and experience has roots in something that has already transpired. However, some-times we have to admit there are developments that are new to us, never envisioned by man, and not directly addressed in Scripture. Computers, DNA sequences, human cloning, genetic engineering of plants and animals—you would be hard-pressed to find ancient wis-dom addressing these issues. And yet Jewish ethicists persist in trying to apply ancient codes of conduct to novel technological issues.

Orthodox rabbis are not troubled by this problem. For many issues, major scriptural principles can be applied—*pikuakh nefesh,* for in-stance, which clearly overrides most other obligations. For some issues, Orthodox rabbis refer to *Tiferet Yisrael,* which states that when the Torah does not specifically prohibit something, then it is permissible.[54]

The Conservative perspective calls for more flexibility and a belief in the evolution of Jewish law. "To address today's medical issues, then, Jewish law needs to be extended considerably," Dorff proclaims. "Contemporary rabbis must take on this challenge if Jewish law is going to be at all relevant to some of the most critical issues of our time. To do so, however, rabbis must face some deeply rooted philo-sophical questions about how to reconcile constancy with change— and indeed how to interpret and apply texts in the first place."[55]

Orthodox rabbis are criticized for trying to force ancient halakhic guidelines to fit novel, modern scientific conundrums. If one accepts the passage "there is nothing new under the sun," it is tempting to try to do this.

From the Orthodox perspective, as expressed by Professor Velvl Greene, who served as Director of the Jakobovits Center for Jewish Medical Ethics in Beersheba, Israel, "In the Jewish world, we don't search for an 'ethical value system' on which to base a medical decision. When faced with an ethical problem, a physician must find a qualified *posek* [rabbinic adjudicator] and pose the question 'This is the medical situation, these are my alternatives, what does the Torah say?' The answer the *posek* provides will have been judged within the same framework as the rest of Jewish life, such as kosher food laws, business dealings, marriage, divorce, and *Shabbat*. And the answer will be an ethical one."[56]

"How can Judaism address contemporary medical realities radically different from those of the past?" challenges Dorff. "The methodological points I have made so far—that we must interpret Judaism's general rules as policies, not inviolable principles; that even general policies must be applied with sensitivity to the contexts of specific cases; and that we must nevertheless retain a legal method in making our decisions—all inform how we should apply Judaism to modern medicine."[57]

And although Dorff rejects the application of ancient precedents to modern questions, he admits that "judges in any legal system must often stretch precedents to make them relevant to new circumstances. Indeed, for a legal system to retain continuity and authority in current decisions, this *must* be done." Dorff further states that the Talmud would have insisted that "each rabbi now take a good look at 'what his own eyes see' to be sure that his or her application of the tradition is deserving of the godly qualities of wisdom and kindness that we ascribe to Jewish law."[58]

The Reform position generally involves rejection of past halakhic precedents. The autonomy sought for by Reform Jews permits decision-making based on tradition, but not bound by tradition. The Reform position articulated by Matthew Maibaum discounts the use of early halakhic rulings, and even basic principles which support those rulings. Dorff rejects this, stating Maibaum's position as "Individual Jews . . . should not feel themselves bound by the traditions' laws or concepts but should rather use the tradition however they wish in arriving at their own decisions."[59] Dorff wonders, "With such an approach . . . how does one rule out anything as being contrary to Judaism?" Dorff disagrees with the Reform view that if a position is not shown to be "immediately and centrally good" according to Maibaum, "then the whole tradition 'is like a fine fossil or an elegant piece of cracked statuary; it is venerable, but is not relevant today.'"[60]

Biotechnology and Bioethics

A recent declaration on bioethics was published by the Biotechnology Industry Organization (BIO). This is of particular significance as that group has served to organize and form consortia of various biotechnology companies. In its Statement of Principles, BIO reaffirms the basic principles of bioethics, declaring: "We respect the power of biotechnology and apply it for the benefit of humankind. . . . We place our highest priority on health, safety and environmental protection in the use of our products. . . . We respect the animals involved in our research and treat them humanely. . . . We adhere to strict informed consent procedures. . . ." In addition, they emphasize the use of biotechnology to "enhance the world's food supply . . . [and] to clean up hazardous waste more efficiently with less disruption of the environment. . . ." The statement also declares, "We oppose the use of biotechnology to develop weapons."[61] These noble concepts of the biotech industry have clearly arisen from general principles of bioethics and have many features in common with principles of Jewish bioethics.

In dealing with cutting-edge technology, ancient standards of conduct do appear to have relevance and value. Whether statuary or statute, fossil or finery, Jewish tradition does speak to us today, and the writings of the past reverberate with relevant messages. It is our challenge to unearth those messages and examine their links to the modern world, addressing how ancient traditions relate to new technologies. With a set of Jewish bioethical principles in hand, we proceed to analyze this brave new world.

CHAPTER 2

Fruit of the Womb

(GEN. 30:2)

"Be fruitful and multiply, fill the earth and subdue it" (Gen. 1:28)

WITH THESE WORDS, the Bible commanded humankind to populate the world, and also directed humankind to use ingenuity, power, and technology to subdue and control the world. Thus, in this verse is a directive regarding reproduction, and, in case natural reproduction does not succeed, a tacit approval for assisted reproduction is assumed. "Be fruitful and multiply" serves as the cornerstone of the obligation and need of Jews to reproduce. Nowadays most rabbis agree that the commandment to populate the world is so important that many modern technological developments for assisting infertile couples may be permitted by Jewish law.

The biblical commandment to have children is the first commandment given to Adam after he was created. A similar directive is given in Isaiah 45:18, which reads: "He did not create the world to be desolate, but rather inhabited." Since Adam was specifically charged to "Be fruitful and multiply," that positive commandment has been interpreted as an obligation on the part of the man to reproduce. The quote from Isaiah, commentators have explained, pertains to both men and women; thus women are included in the obligation to fill the world.

It is clear that the Scripture has commanded all people to procreate, and this directive is so critical that Torah scholars agree it could be accomplished by natural or artificial means. The challenges of assisted reproductive technologies are to sort out the complex relationships created by artificial reproductive processes and to determine where to draw the line in terms of what techniques are ethical and permissible, which advances are questionable and which are unacceptable.

The implied flexibility of the Torah regarding assisted reproduction should not be surprising. After all, three out of four biblical Matriarchs suffered from infertility. The suffering of Sarah, Rebecca and Rachel due to infertility are documented in detail in Genesis, and provide much of the human drama in the relationships between the Matriarchs and their husbands, and the Matriarchs and God. The three Matriarchs dealt in different ways with their tragic circumstances. Sarah bitterly resigned herself to not having children, and even laughed a cynical laugh—"And Sarah laughed within herself" (Gen. 18:12)—when presented with the possibility of conception at an advanced age. Rebecca was more positive; she asked Isaac to intervene on her behalf. "And Isaac entreated the Lord for his wife because she was barren" (Gen. 25:21), and Isaac's prayers were answered. Rachel resorted to more desperate measures. She declared to Jacob: "Give me children, otherwise I am dead" (Gen. 30:1). The commentator Rashi explains that this statement signifies that a childless person is accounted as dead.

Rachel's next act was even more desperate. Reuven, the firstborn son of Leah, returned from the field with some plants called "*dudaim*" (Gen. 30:14). The biblical commentator Nachmanides suggested that these plants were herbs that promoted conception. Reuven presented them to his mother for her use. Rachel observed this and begged her sister for the plants. Then she made a deal: in return for the *dudaim* she would allow Leah to spend one night with Jacob. Ironically, Leah's fifth son was born as a result of this deal. And because of Rachel's desperation this "assisted reproduction" achieved its goal for her as well. Rachel was finally "remembered" by God, and she conceived and bore Joseph. She then states: "God has taken away my disgrace" (Gen. 30:23).

Infertility was not only a painful and tragic experience for the Matriarchs. It continues to afflict many Jewish couples. Although the biblical instances of infertility appeared to be cases of female infertility, the cause of infertility among Orthodox Jewish couples today may be more commonly due to male factors. Dr. Vincent Brandeis, who runs fertility centers in New York State, has estimated that in 60 percent of Orthodox couples infertility is due to problems with sperm quality.[1] This contrasts with the general population in which estimates of male factor infertility range from 40 to 50 percent.[2] "The husband has been found to be a significant factor in about 30 percent of these cases of infertility and to play an important role in another 20 percent."[3] Why would there be such a discrepancy in the rate of male factor infertility in Orthodox couples compared to the general population? According to Brandeis, the reason for this difference may have to do with the low incidence of pelvic inflammatory disease among Orthodox women, who are generally sexually inactive before marriage. Pelvic inflammatory disease increases the rate of female infertility in the general population so that it slightly exceeds male infertility.

One aggadic notion found in the Talmud regarding reproduction in many ways reflects the views of those times. "Our Rabbis taught: There are three partners in man: The Holy One, blessed be he, the father and the mother. The father supplies the semen, the white substance, out of which are formed the child's bones, the sinews, the nails, the brain and the white of the eye. The mother supplies semen, the red substance, out of which are formed the skin, flesh, hair, blood and the black of the eye. God provides the spirit, the soul, the beauty of the features, eyesight, the power of hearing, ability to speak and walk, understanding and intelligence."[4] Although the Talmud was unable to explain the scientific mechanisms of inheritance of traits, it was enlightened in assigning a role to the female "semen." Many other scholars of the time, on the other hand, believed that the female was simply an incubator in which the male seed grew into a child. Even the Torah recognized the presence of a female seed with the passage: "If a woman emits a seed" (Lev. 12:2).

Modern Concepts and Practice

Today, of course, Jewish authorities approach questions of infertility and reproduction from the perspective of scientific understanding as well as from halakhic precedent. The basic principles of genetics are well documented. There are many thousands of genes—current estimates are in the range of 40,000—inherited by a child that control the physical attributes of the child. For most traits a child inherits two copies of each gene, one from the mother and one from the father. The individual copies of each gene can interact with each other. A copy from one parent may be dominant over the other and be preferentially expressed in the child. Or the two copies may work together to produce a combination or blended trait.

Assisted reproductive technology that is available today is enabling many infertile couples to fulfill the biblical commandment to "be fruitful and multiply" (Gen. 1:28). Technological advances have led to the development of in vitro fertilization (IVF), where the sperm from the father and the egg from the mother are mixed together in a petri dish in the laboratory, and the sperm is allowed to fertilize the egg, producing a "test-tube baby." The fertilized egg is then returned to the biological mother's womb where it develops, and nine months later the baby is born. In IVF, conception takes place outside of the body. This technique overcomes the problem of scarred, damaged, or blocked fallopian tubes (which prevent the sperm from reaching the eggs, and the eggs from reaching the uterus). It also allows men with low sperm counts to conceive, as sperm samples can be concentrated and deposited adjacent to the ripe eggs. Only small numbers of viable sperm are needed for successful fertilization in a petri dish.

A variation on this theme is intra-cytoplasmic sperm injection, or ICSI. With this technique a man who produces no sperm at all in his ejaculate can become a father. Pieces of tissue from the testicle can be used. A few sperm cells are isolated from the ejaculate or directly from the testis, and mechanically injected, one by one, into individual eggs (see chapter 3). Another variation involves a process called Assisted Hatching. In Assisted Hatching a small opening is made in the clear zona, or "shell" around the egg. The process is done to allow the fertilized egg to emerge properly, as this can assist it to implant into the lining of the uterus. In the late 1990s, Israeli doctors developed a method that uses laser energy to open a window in the zona. Other methods, one of which involves mechanically tearing the zona with a sharp needle, appear to increase the incidence of twinning.

Halakhic Implications

How "kosher" are these techniques in terms of Jewish law (halakha)? It is generally agreed by rabbinic authorities that IVF and related techniques are acceptable for Jewish couples when the husband's sperm and the wife's eggs are used. But even the use of IVF by a married couple has resulted in some surprising rabbinic reactions. For instance, Rabbi Eliezer Yehudah Waldenberg rules that a child conceived outside of the womb has *no* halakhic parents.[5] This ruling has been sharply criticized by a Central Conference of American Rabbis (or CCAR, which provides Reform Jewish policy) responsum.[6] Waldenberg cited Maimonides' statement that "human organs cannot exist separately from the body and still be regarded as fully human."[7] Extrapolating from that principle, a human egg in a petri dish ceases to be human, hence would no longer retain a relationship with its biological parents—the egg or sperm donor. In fact, argues Waldenberg, the gestational mother also would not incur the status of motherhood. "No *yichus* (familial relationship) is possible outside the womb of a Jewish woman."[8]

Most rabbinic sources disagree with the Waldenberg ruling, and ascribe full parental rights and relationships between couples who conceive using IVF and their offspring. However, there are some halakhic issues that can complicate husband/wife IVF.

Multiple Pregnancies

When more than one fertilized egg is implanted into the woman, this may result in a multiple pregnancy. When there are three or more fetuses growing in the womb, this results in a high-risk pregnancy, and fetal reduction, or selectively eliminating one or more of the fetuses, may be recommended. Is this halakhically permissible? Ending the life of a fetus is not considered murder by halakhic definition, but it is not permissible either. This would only be permitted if the doctor has determined that some fetuses must be eliminated or they will all die. In certain cases, the fetus can be considered as an aggressor or a pursuer *(rodef)* who threatens the life of the mother. In those instances the fetus may even be dismembered, limb by limb, because the mother's life takes precedent over the life of the fetus.[9] Whether the fetus can be seen as an aggressor vis-à-vis the other fetuses is another question. The decision regarding fetal reduction is a very sensitive one and must be made by the doctor.

Rabbi Elliot Dorff addresses this question. "Since it is dangerous to both the mother and the fetuses if she carries more than three, this results in the need to abort some of them to reduce the number left to a maximum of three." He strongly recommends that this situation be avoided by adopting the practice (which is common in European IVF clinics) of implanting no more than three embryos per cycle.[10]

Extra Eggs and Embryos

When IVF is performed, the woman is stimulated by hormone treatment so her ovaries can produce up to twenty eggs per cycle. The eggs are harvested and fertilized, but only three or four can be used in that cycle. The rest can be preserved by freezing. How does Jewish law address the issue of extra embryos? The fate of extra embryos could include:

A. Use of them by the original couple to establish future pregnancies. They can be implanted at some future time in the original mother to establish additional pregnancies. Rabbis affirm this use by the couple who produced the embryos, since it improves their chances for conception, which helps fulfill the duty of procreation. Since long periods of time can elapse between freezing and future use, siblings conceived at the same time may be born years apart. It is even possible that a sibling conceived years later may be born before the embryos that were frozen away in IVF. One possible question that could arise has to do with inheritance. However, since birth order determines birthright, and not time of conception (see, for instance, the story of Esau and Jacob, Gen. 25:24–26), this should not present a halakhic problem.

B. The extra embryos may be destroyed. This is halakhically permissible according to most rabbis if it is done passively, for instance, by letting the embryos thaw out and die on their own. (More details on the status of embryos are presented in chapter 4.)

C. The extra embryos can be used for research. (Further discussion on this issue appears in chapter 4.)

D. The extra embryos can be donated to another infertile couple. This can pose numerous halakhic problems (see below).

Meticulous Supervision of IVF Labs

Although most IVF labs are reputable and try to be meticulous in keeping track of the sperm, eggs, and embryos of each couple, over the years some mistakes have been made. Having worked in an IVF lab, I can attest to the fact that all eggs and embryos look alike. Scrupulous attention to labeling test tubes and petri dishes, and careful record-keeping are of paramount importance.

In a recent case in New Jersey, a white woman who participated in IVF gave birth to two boys, one white and one black. The black one turned out to be the biological son of a black couple who were having IVF at the same time in the same clinic. Apparently some of their embryos were implanted into the white mother's uterus. After a protracted lawsuit, eventually the black child was returned to his genetic parents, and all legal ties to the white parents were severed. That case is an example of a situation where an error could be readily discerned because of the difference in appearance due to race. It is reasonable to assume that other errors occur in IVF labs that may go undetected; and other parents may be carrying and delivering babies who are not their own.

Even worse than inadvertent errors are the cases of deliberate tampering with sperm, eggs, and embryos that have been uncovered in unscrupulous fertility labs. Dr. Ricardo Asch from the University of California, Irvine, was charged with using extra eggs from infertile couples without the permission of the patients. The eggs were used to establish pregnancies in other women who needed donor eggs. At least fifty-three lawsuits have been filed by patients whose eggs were used without their consent. One couple who had extra embryos frozen back in 1991 learned years later that some of those embryos had been transferred to another woman who gave birth to twins. Some of these cases have developed into full-blown custody suits between the genetic parents and the gestational mother.

Another shocking case involved Dr. Cecil Jacobson, an infertility doctor from Alexandria, Virginia. In 1992 he was convicted of artificially inseminating hundreds of his patients with his own sperm—while telling them he was using donor sperm. He is said to have fathered dozens of children as a result of these deceitful practices.

Since parentage is of vital concern, some Orthodox rabbis recommend the use of trained supervisors during IVF procedures involving Jewish couples. The Puah Institute, which is an Israeli based organization whose mission is exemplified by their motto "Fertility and Medicine in Accordance with Halacha," provides such supervisors for many IVF programs. The name of the institute is derived from the biblical midwives Shifrah and Puah, who were ordered by Pharaoh in Egypt to kill all baby boys born to the Hebrews. They did not comply, and were rewarded for this by a subsequent population explosion in the Israelite camp (Exod. 1:15–21). According to the Talmud, Shifrah and Puah were pseudonyms for Yocheved and Miriam, mother and sister of Moses, respectively.[11]

One of the activities of the Puah Institute is a "training program for Haredi [ultra-Orthodox Israeli] women to become supervisors, monitoring all stages of the total IVF procedure."[12] The female supervisors, called *Mashgikhot,* accompany the sperm and egg and ensure that sperm, eggs, and embryos are meticulously inventoried and tracked. In the past five years alone, reports the institute's Rabbi Gideon Weitzman, the *Mashgikhot* discovered and prevented eighteen errors, where the egg was going to be put back into the wrong woman. And when the Puah Institute started to offer its services in a hospital in France they discovered—and prevented—a mistake on the very first day of supervision. Currently, the Puah Institute offers supervisory service at dozens of hospitals in Israel as well as a hospital in Paris, Boca Raton Hospital in Florida, and several hospitals in New York. In addition, other organizations are also getting involved. Star-K Certification of Baltimore, Maryland, and a New York City clinic reportedly developed such an arrangement, and Shaarei Zedek Hospital in Jerusalem has its own *Mashgikhot.*[13]

Donor Eggs

When an egg donor provides an egg for an infertile couple, the recipient, usually a sterile woman who cannot produce eggs, serves as the gestational and birth mother and she gives birth to and raises the baby as her own. In this case there are two categories of motherhood: a genetic mother and a gestational/birth mother. These—now separate—functions can be performed by two different people, who may or may not be related to each other and may or may not have any connection with each other (other than their individual contributions to producing and raising the child).

Weitzman, from the Puah Institute, reported that there are children in the *haredi* community who are born from egg donors. Apparently there is a controversy regarding whether it is preferred to use eggs from Jewish or non-Jewish donors. When an egg or an embryo from a Jewish woman is donated to another, unrelated couple, the "adopted" child may inadvertently marry his/her genetic sibling, resulting in incest.[14]

"A sexual relationship between paternal half-siblings all of whose parents are Jewish is considered halakhically incestuous. However, in contrast to most contemporary secular societies, halakha sees no incestuous relationship between half-siblings who have a common non-Jewish father (irrespective of the religion of the mothers). On the other hand, a relationship between half-siblings who have a common mother (Jewish or not) would be incestuous irrespective of the religion of the fathers," explains Dr. Richard Grazi. He continues, "There is a general reluctance to allow egg donation based on the unsettled questions of religion and maternity (including the problem of future incest). . . ."[15]

Frozen embryos have typically been donated to unrelated couples; but there also have been cases of grandmothers gestating their own grandchildren, and aunts carrying their own nieces and nephews to term. According to Grazi, "Most ovum transfer patients surveyed had not only considered using a sister, but 61 percent had secured such an agreement." (It is interesting to note that only 11 percent of couples using donor insemination would consider using a brother as their donor—"and none had actually asked one to participate."[16]) Egg or embryo donation to close relatives can also cause halakhic concerns, and could result in illicit relationships. In halakhic terms, if you give birth to your own genetic grandson, does he become your son? Is he then also the half-brother of your daughter (his real genetic mother)?

There are rabbinic authorities who reject outright the idea of using donor eggs. Others believe that a woman may receive donor eggs as long as her husband has consented. The answer to the question of who is the mother is extremely complicated. This is certainly a critical question as it affects the status and identity of the baby. According to traditional Judaism, the status of "Who is a Jew?" is determined by whether or not the mother is Jewish. In the case where the genetic mother and the gestational mother are the same person, then the issue is clear. What happens when the genetic mother is a different person from the gestational mother? Which mother is considered the mother for the halakhic decision on religious status? If the genetic mother is not Jewish and the gestational mother is, what is the status of that infant?

The most extreme view of Waldenberg, as stated above, is that once the egg is removed from a woman's body, all familial relationships are severed. Another stringent ruling, of Rabbi Moshe Heinemann, rabbinic administrator of Star-K Kosher Certification, states unequivocally that if the egg is from a non-Jewish woman, then the baby is not Jewish. In this ruling, when a donor egg is used, the birth mother is not considered the halakhic mother.[17]

Rabbi Moshe Tendler writes: "the contributions of the gestational mother are quite consequential."[18] In fact, many halakhic authorities regard the birth mother, rather than the egg donor, as having maternal status. Rabbi David Feldman concurs, offering the argument that "[n]ine months of responsible nurture is a far greater 'gift' than the surrender of an ovum, entitling the host far more clearly to acknowledgment of parenthood."[19]

On many issues, halakha relies on what can be readily observed with the naked eye. For instance, microscopic or small amounts of nonkosher contaminants in kosher foods do not necessarily render the food nonkosher. Thus, one might conclude that the decision on maternity may be based on which mother gives birth (an action that is incontrovertible and readily proven), rather than which mother provided the egg (a microscopic contribution, albeit a critical one). However, it must be emphasized that just as trace quantities of food additives can render food nonkosher if they change the nature of the food, the microscopic nature of the egg does not negate its importance. Feldman explains, "Just as the atom may be unseeable, but the atomic bomb must be reckoned with, so halakhah reckons with the effects of the unseen. Hence questions of fertilization, and a myriad of biological as well as food-composition processes, are very serious halakhic concerns."[20]

Feldman therefore maintains that "the genetic, as opposed to the gestational, contribution is the dynamic one, giving the all-important genetic code. Also, the donated zygote can be said to be a nascent being, with the identity of the mother already intact."[21]

In addition, considering the important role *yichus,* or inherited status, plays in some Jewish circles, genetic status would be of paramount importance, thus, perhaps the mother who provided the egg should determine Jewish status.

Rabbi Elliot Dorff also presents two sides of the argument regarding whether the egg donor or the gestational mother is considered the halakhic mother. He explains that Judaism identifies the firstborn son—he who needs to be redeemed from temple service in the Pidyon Ha-ben ceremony—as the child who is "*petter rekhem*" (the first issue of every womb) (Exod. 13:2, 12, 15). Based on this, "it may be that the bearing woman, rather than the egg donor, should be defined as the child's mother," reasons Dorff. "On the other hand, one might argue that there should be a parallelism between the identity of the father and that of the mother, and since Jewish law defines the sperm donor as the father, the egg donor should likewise be seen as the mother."[22]

The Committee on Jewish Law and Standards of the Conservative movement has ruled that the biblical phrase *petter rekhem* (the opening of the womb) determines who the mother is. As described by Dorff, "It is thus the bearing mother who determines the Jewish identity of the child; if she is Jewish, the child is Jewish, regardless of the source of the egg used in the child's conception; if she is not Jewish . . . the child is not Jewish by birth and must undergo the rites of conversion to become Jewish."[23]

The CCAR Responsum on "In Vitro Fertilization and the Status of the Embryo" emphasizes the relationship of the genetic parents to the offspring, but concludes that "[t]he embryo may be offered to another couple. The child will be the biological offspring of the man and woman who donated the sperm and the egg. Those who raise the child are his or her 'ultimate' and 'real' parents."[24]

Rabbi J. David Bleich offers an unconventional scenario regarding maternal status. "Perhaps two maternal relationships may exist simultaneously, just as maternal and paternal relationships exist at one and the same time. The child would then, in effect, have two 'mothers,' the donor-mother and the host-mother." His reasoning is based on "the seed of the father" determining fatherhood; by extrapolation, the "seed" of the mother (the ovum) determines motherhood. However, the gestational mother also retains rights. Since a transplanted organ takes on the status of the body of the recipient, a donated ovum would also become part of the gestational mother's body—she acquires ownership of that ovum; when she gives birth, it is her own child. Thus, the ovum donor and the gestational mother may both be considered halakhic mothers.[25] In fact, according to Feldman, Israeli hospitals have been known to list both egg donor and gestational mothers on the birth certificate.[26] According to Bleich, under this reasoning—involving dual maternity—if either woman is a non-Jew, then the child would have to undergo conversion to be considered Jewish.[27]

Feldman expands on the complications involved. "*Kibbud em* ('honor thy . . . mother') would be owed to both, and children born to both would be siblings forbidden in marriage."[28] The unfortunate child might, thus, theoretically, have to answer to two mothers. And under some circumstances, that could mean two *Jewish* mothers.

A recently developed technique to combat female infertility may confound the situation of maternal identity even further. Eggs produced by some women exhibit a very low rate of successful fertilization in IVF procedures. It turns out that some factor in the cytoplasm of the egg may be lacking and, as a result, fertilization does not occur. The cytoplasm is the portion of the egg surrounding the nucleus. The egg nucleus, which is the main genetic compartment, holds 99 percent of the genes of the mother. The cytoplasm, on the other hand, contains the remainder, or about 1 percent of the genetic material (in cell structures called mitochondria). A procedure developed in 1997 to stimulate "reluctant eggs" involves injecting a small amount of cytoplasm from a donor—a fertile woman whose eggs easily fertilize—into the eggs of a recipient whose eggs are not otherwise able to be fertilized. The cytoplasm apparently contains some factors involved in activating sperm after fertilization, thus injection of donor cytoplasm appears to help otherwise unfertilizable eggs.

The procedure increases chance of fertilization for some IVF participants. However, one consequence of this procedure is that with the small amount of cytoplasm from the donor, some mitochondria are also transferred to the recipient egg. So the recipient egg is also receiving a small amount of mitochondrial DNA, that is, a number of genes from the donor cytoplasm. The end result is that babies born from this fertility treatment have genes from two women (and one man). Those children will inherit most of their genetic complement from their father and mother (the sperm and the egg); in addition, they will inherit mitochodrial genes from the cytoplasm donor. What would the halakhic ramifications of a child with three genetic parents be? If both members of a lesbian couple want a genetic link to their child, could they elect this procedure? Would they then both be legally considered to be the mothers of that child? What are the societal implications? This is not a hypothetical situation—these children really do exist and are quite normal; they simply have extra mitochondrial genes from another source. This type of technological development will certainly stimulate discussion as these issues will have to be resolved from a variety of standpoints.

In the interests of fulfilling the commandment to populate the Earth, most rabbinic authorities are fairly accepting of artificial reproductive technologies. However, with some of the available technologies, questions regarding the definitions of parenthood abound. Approaches to infertility treatment such as the use of donor eggs, donor cytoplasm, and surrogate wombs raise issues that are still under discussion.

One other interesting spin on maternal fertility is provided by a midrash stating that part of the happiness of Paradise will be that "a woman will there give birth to a child every day."[29] No doubt that midrash was written by a man.

CHAPTER 3

Be Fruitful and Multiply: Male Infertility

"A man must not abstain from *fruitfulness and increase* unless he has already children. The School of Shammai say, Two sons; but the School of Hillel say, A son and a daughter, because it says 'Male and female He created them.' . . . It is the man who is commanded regarding the propagation of the human race but not the woman."[1]
—MISHNAH

PATERNITY IS OF paramount importance in Judaism. The laws of inheritance depend on identifying the legitimate offspring of a man. The determination of tribal affiliation—Kohen, Levite, or Israelite—is through the father as well. In Orthodox and Conservative Jewish practice Jewish status is transmitted through the maternal line. Reform Jews, on the other hand, permit Jewish status to be determined through maternal or paternal lineage.

Assisted reproductive technologies developed in the 1990s have had an impact on the role of the Jewish male in conception and have raised interesting issues with regard to the male contribution to conception. The Scripture has touched upon some unusual points related to reproduction. Modern rabbinic references grapple with these issues realizing that, according to Rabbi Gideon Weitzman of the Puah Institute in Jerusalem, sometimes "there's no answer in the *halakhic* material."[2] So rabbis have to look for halakhic precedent for scenarios that never previously were possible. When considering the male contribution to reproduction, the production of and status of sperm,

seminal fluid and ejaculate are important, as is the anatomy and phys-
iology of the male organs.

The Role of the Sperm

Every somatic (nonsex) cell in the body carries twenty-three *pairs* of
chromosomes, for a total of forty-six. Twenty-three are paternal, or
inherited from the father, and twenty-three are maternal, or inherited
from the mother. These cells are considered "diploid" because they
have two copies of each chromosome. The sex-cells are different.
When spermatozoa, or sperm, are produced in the testis they go
through a special process called meiosis in which they divide and produce cells carrying only twenty-three chromosomes. These "haploid"
cells have only one copy of each chromosome, instead of a pair. A similar process occurs when eggs are made. A sperm with twenty-three
chromosomes fertilizes an egg with twenty-three chromosomes to re-
establish the diploid number of forty-six chromosomes in the new embryo. So every sperm cell carries half the genetic complement to produce a new embryo. The job of the sperm is to carry those genes to the
egg and deliver them inside.

One of the easiest tests to perform on an infertile couple is the
sperm motility test. The man ejaculates into a sterile container; the
semen coagulates and becomes jelly-like immediately after ejaculation. Within minutes, it liquefies again and the sperm can be sus-
pended in nutrient medium for subsequent analysis. The sperm are
evaluated microscopically with regard to number, appearance and
motility—or vigorous movement. Although for fertile men there is a
wide range of variability in sperm samples, even within normal
ranges, men who produce fewer sperm may take longer to conceive.
Specimens in which there are few or no sperm swimming vigorously
would be classified as subfertile or sterile, depending on how sluggish
the sperm are. A normal sperm count is defined as having a volume
of at least one milliliter, a sperm count of at least 20 million sperm
per milliliter, and at least 50 percent of them should be vigorously
moving.[3]

In the case of low sperm count, if the sperm that are present are
strong swimmers, it may be possible for the sperm to penetrate eggs on
their own. Because there is dramatic attrition as sperm make their way

from the vagina to the fallopian tubes where an egg may be waiting, even vigorous swimmers may need to be concentrated in a smaller volume and deposited closer to the egg. For instance, by placing sperm into the uterus (intrauterine insemination), the cervical barrier is bypassed and a higher percentage of sperm will be able to reach the egg. In fact, in some cases, the cervix produces mucous that is inhospitable to sperm, which may present an insurmountable barrier for some sperm.

When a man's semen has a very low sperm count, the viable sperm can be concentrated and placed in the area immediately surrounding the egg, either in a petri dish or in the fallopian tube. In in vitro fertilization (IVF) sperm and eggs are placed together in a small petri dish. Gamete intra-fallopian transfer (GIFT) involves the placement of egg and sperm together in the fallopian tube. However, if the quality of the sperm is poor, that is to say, very few or none are vigorous swimmers, then these techniques will not help. To effect conception in this case, sperm must be physically forced through the membrane of the egg. The technique, called intra-cytoplasmic sperm injection, or ICSI, where a sperm is physically forced into an egg, has only been available since 1992.

It is amazing that conception ever occurs, considering that the microscopic human egg presents a formidable challenge for the sperm. Each egg is a living cell bounded by a fatty covering called the cell membrane. Surrounding the egg's membrane is a thicker, transparent shell called the zona pellucida (literally, clear zone). Around the egg and its shell is a thick cloud of cells, called cumulus cells (which look a bit like cumulus clouds in the sky). The challenge for a sperm cell is to disperse the cloud of cells, burrow through the zona, and fuse with the cell membrane, depositing the sperm's chromosomes inside. Each sperm is equipped with a reservoir of enzymes that can loosen and disperse the cellular cloud and can assist the sperm in digging into its target. Once one sperm gets all the way in, a chemical reaction occurs changing the nature of the egg membrane. It becomes impermeable to most other sperm, essentially allowing only one sperm in and then "locking the door." That chemical reaction is called the "block to polyspermy." It is a critical process since it prevents more than one sperm from fertilizing the egg. If two sperm cells were to penetrate the egg the result is a genetic disaster—a human embryo with sixty-nine chromosomes rather than the normal forty-six.

In talmudic times the possibility that more than one sperm could produce a child was considered and believed to occur. In fact, the Jerusalem Talmud discusses Superfecundation—when a woman is pregnant at one time from two men. The case of a woman who is carrying a child fathered by more than one man is described by Julius Preuss. "Since conception occurs within three days of cohabitation, if the woman copulates with another man during this time period, then mixing of the sperm may occur and the child may, in fact, have two fathers." In his 1911 book *Biblical and Talmudic Medicine,* Preuss mentioned that embryologists of *his day* also believed that several sperm can penetrate a single egg.[4]

We know from research on in vitro fertilization that occasionally this does happen—more than one sperm enters an egg. However, usually only one man provides sperm for an IVF cycle, so superfecundation involving two fathers would not be typical. As an embryologist in Mount Sinai's IVF lab in New York City, I observed human eggs shortly after fertilization. In normal fertilization, about twelve hours after a single sperm gets in, the twenty-three maternal chromosomes and the twenty-three paternal chromosomes are observed in two separate round structures, called pronuclei. After about twenty-four hours, the forty-six chromosomes (twenty-three pairs) mix and the embryo begins to divide into two, four, eight cells, and so on. However, in eggs that had been penetrated by more than one sperm, shortly after fertilization three or more pronuclei are observed. If two sperm enter, for instance, two paternal and one maternal pronuclei are present. Such an embryo is called triploid, since it has three copies of each of the twenty-three chromosomes (a total of sixty-nine), instead of the usual pair (totaling forty-six). A triploid embryo arises because there was something seriously wrong with the egg, and it lost its capacity to block a second sperm from entering. In fact, sometimes faulty eggs do not fertilize at all, sometimes they fertilize abnormally and fail to develop, and sometimes they develop for a few days and then die.[5]

The proper development of a human depends not only on the types of genes present, but also on the amounts of genetic material. If a fertilized egg ends up with even one extra chromosome (a total of forty-seven, which is called trisomy), the genetic imbalance can create instability and gene dysfunction. Most embryos that carry forty-seven chromosomes do not survive to birth (one exception is Down syndrome, in which the baby born carries an extra chromosome # 21—and exhibits a multitude of birth defects including mental retardation). Thus it stands to reason that an embryo with twenty-three extra chromosomes (triploid) would be severely abnormal.

If a triploid embryo resulted from sperm of two different fathers, it would be an example of superfecundation. However, triploid embryos are so genetically aberrant most do not develop for more than a few days or weeks after conception. In the rare instance that a triploid baby is born, such an infant, born with severe defects, does not survive long. Thus, ancient beliefs in superfecundation do have some biological basis, in that it is possible for such a conception to occur. However, we now know that babies generated by superfecundation would not survive.

Of course, if a woman ovulates more than one egg in a given cycle, two or more babies can arise from the fertilized eggs; and babies from two (or more) different fathers could occur in one pregnancy. This could occur if she has relations with two men around the time she is ovulating and sperm from more than one man are present.

Oddly enough, this can also occur as a result of infertility treatment in some clinics. In some infertility clinics, when the husband's sperm is judged subfertile, the couple may be advised to use a sperm donor. Some doctors suggest that the husband's sperm be mixed with the donor sperm before insemination. The reasoning is that, by mixing vigorous donor sperm with the husband's sluggish sperm, it may enhance the husband's (albeit slim) chance of becoming a father. The paternity of the offspring may subsequently remain a mystery, if the couple so desires. Or they can perform a genetic test to determine paternity. In this scenario, if more than one egg was ovulated by the woman, there is a chance of a multiple birth producing children with different fathers.

Male Infertility

The paternal contribution to infertility was considered by talmudic sources. Mishnah *Yevamoth* states: "If one took a wife and lived with her ten years and she bare no child, he is not permitted to abstain. If he divorced her, she is permitted to be married to another." The male contribution to infertility was not ruled out; it was not assumed that she was the one who caused the infertility; thus, she is permitted to marry another man. Only after her second union is barren for ten years is she considered a sterile person. At that point, after two ten-year periods of no conception with two separate men, she must be divorced and "may not marry again as she is considered a confirmed sterile person."[6]

There are several other instances where biblical and talmudic sources consider the possibility that a man can be infertile. "This shall be your reward when you harken to these ordinances," states Deuteronomy (7:12). The text continues, "there will be no infertile male or infertile female among you nor among your animals" (Deut. 7:14). Julius Preuss also cites the story of Manoah and Hannah (Judg. 13:2), who suffered infertility until their son, Samson, was born. According to legend, this couple "disagreed among themselves as to who was the infertile partner of their childless marriage."[7] In addition, the Talmud discusses the following unusual observation: "Rabbi Acha bar Jacob stated that many scholars became impotent because of the long discourses of Rabbi Huna."[8] Apparently the young men were forced to sit for such long periods of time that it reportedly affected their ability to perform sexually.

The injunction to procreate is so paramount that, according to Mishnah *Nedarim,* if a woman stated "Heaven is between you and me" (meaning, only heaven knows that you are not able to perform coitus), then the husband could be forced by the court to divorce his wife and pay her *ketubah* (the value of her marriage contract).[9] Men who had penile malformations and were physically unable to perform intercourse could be prohibited from marriage with Israelite women. The Talmud describes different abnormalities of the penis and the testicles that might lead to infertility in men. Deuteronomy 23:2 commands that a *keruth shofkha*—a man whose penis is cut off or cut into—"shall not enter the congregation of the Lord." Since such a man still has sexual desires, however, the Talmud permits him to marry a female convert or a freed female slave.[10]

In Talmud *Bekhoroth,* the anatomy of the penis is described, according to the prevailing view of the time. This description—although inaccurate—also attempts to explain some cases of infertility. "There are two channels in the penis of man, one for the discharge of urine and the other for sperm. Between the two, there is a separating membrane as thin as an onion peel. If a person does not respond to the call of nature (i.e., to urinate) and one of these ducts perforates into the other, that man becomes infertile."[11] In actuality, there is only one channel, the urethra, which runs through the penis and carries either urine or semen. When semen are emitted, urine is blocked from entering the urethra.

Handling Sperm

Although no microscopes existed in biblical or talmudic times to reveal the secrets of male fertility, semen was understood to play a key role in conception and was assigned a special status. Practices concerning seminal emissions were complex. "According to Biblical command, if a man emits sperm, he should wash his entire body in water and he remains ritually unclean until evening. Every garment and every skin upon which sperm has come should be washed in water and is unclean until evening (Lev. 15:16–17). In times of war, such a man should not return to camp until he has bathed (Deut. 23:11)."[12] The topic of emission of seed was of great interest and importance and was further discussed in Mishnah *Mikvaoth*.[13]

Ancient observations and conjectures on conception and sexual performance were provided by several talmudic sources. "Ejaculation only impregnates if 'the entire body feels it' [orgasm]. Mar Samuel said: 'sperm which doesn't shoot forth like an arrow doesn't fertilize'"[14] On the other hand, reported the Talmud, "through energetic diversion, one can avoid ejaculation; one should bore his fingernails into the ground until his desire dies out."[15]

The story of Onan and Tamar (Gen. 38:9–10) involves Onan spilling his seed rather than impregnating Tamar, his dead brother's wife. "What he did was evil in the eyes of God, and He caused him to die also." Since there is a biblical admonition regarding the "spilling of seed," some rabbis insist that the husband may not ejaculate to provide a specimen for fertility testing. However, since the intention of the procurement of semen is specifically to enhance procreation and the semen is not being wasted, ejaculation to produce the semen may indeed be permissible. Most rabbis today accept the testing of sperm as a means to achieve the lofty goal of conception. "Of paramount importance is purpose," writes Dr. Yoel Jakobovits. "When the intention is procreation, either directly through artificial insemination and in vitro fertilization or indirectly when evaluating male infertility, there is significant room for leniency within the halakhic guidelines."[16]

Rabbi Moshe Feinstein, a major halakhic adjudicator of the twentieth century, supported the use of sperm analysis for fertility testing. Dr. Fred Rosner reported, "It is permitted by most rabbis to obtain sperm from the husband both for analysis and for insemination, but difference of opinion exists as to the method to be used in its procurement. Masturbation should be avoided if at all possible, and *coitus interruptus* or the use of a condom seem to be the preferred methods."[17]

Sterility

When a man produces no motile or viable sperm, is there any way he can biologically father a child? There are options now, thanks to astounding advances in reproductive biology, such as ICSI (intra-cytoplasmic sperm injection). ICSI provides a mechanism to force an immotile sperm cell, and even an immature sperm, into an egg. Although just a few decades ago only 50 percent of infertility cases were addressable, now, according to Rabbi Gideon Weitzman of the Puah Institute, "for almost 95 percent of cases today, we have an answer."[18] And ICSI is only one significant reason for this increase in success in the treatment of infertility. Men who had previously no hope of becoming fathers are now conceiving their own biological children.

In the early 1980s, I was involved in one of the first projects on sperm micro-injection. In these early experiments we were attempting to fertilize mouse eggs by sliding mouse sperm between the clear shell of the zona pellucida and the egg membrane. We were concerned that if the sperm were shoved all the way in through the cell membrane, it would disrupt and destroy the integrity of the egg, essentially killing it. Other ingenious approaches that were attempted by investigators included chemically dissolving the zona and physically tearing the zona, to make a window for sperm to more easily reach the egg membrane.

These experiments were repetitive and tedious, but the anticipation of advancement in the field made this an exciting endeavor. All work was done using a special microscope equipped with special tools to pick up and handle sperm and eggs, one at a time. These microscopic tools were produced from slender glass tubes that were heated and drawn out to render them into finely pointed instruments. Some of the tiny glass tools were further heated to produce smooth, polished suction devices (micropipets). Micropipets were used to draw up and hold each egg in place for microsurgery; this was somewhat analogous to using a vacuum cleaner to hold a bowling ball—as seen in Oreck's vacuum cleaner advertisements—albeit on a very tiny scale.

Other glass tubes were sharpened to fine beveled points. Those micro-needles (which had to be ground at a specific angle) were used to pick up sperm, one at a time, and puncture the zona, carrying the sperm in. The micro-needle would draw each sperm—tail and all—through the eggshell as a needle draws a thread through cloth. After practicing on hundreds of mouse eggs (many of which exploded on impact), the technique began to work in our lab. The only problem was, putting sperm under the zona did not fertilize the egg. I observed, however, that when sperm were accidentally shoved all the way through the membrane, an occasional egg did survive and displayed signs of fertilization and early development.[19]

A modification of this technique was later developed for use with human eggs and sperm. Fortunately, human sperm are smaller than mouse sperm, hence when they are micro-injected they cause much less damage; many more eggs survive this process and are fertilized. That is important, since human eggs are much harder to come by (and much more precious) than mouse eggs.

Our ability to surgically manipulate eggs, sperm, and embryos has revolutionized infertility treatments. Thanks to ICSI, some males who do not produce any viable sperm are able to become biological fathers. Rabbinic reaction has been positive; considering that this technique now makes it possible for men to fulfill the obligation to procreate, that reaction is not surprising.

However, given the greater implications of sperm injection—that humans, rather than God, or chance, determine which sperm gets in—it may be surprising that rabbinic reaction is so positive. The Catholic church has made it clear that they consider human control over procreation to be overstepping the bounds of what is permissible.[20]

Does ICSI Raise Ethical Issues?

"There was a big backlash against ICSI," stated Rabbi Gideon Weitzman, "especially in [the United States], because we don't know with

ICSI if we're transferring diseases which otherwise would not have been transferred."[21] In other words, we are enabling children to be born who would never have been conceived, because in this scenario humans are propagating offspring who would not be considered genetically select. Boys conceived using ICSI might be in danger of being sterile like their fathers if the father's infertility is due to genetic causes. Some fertility centers report higher incidents of chromosome abnormalities in ICSI babies, while others do not find any difference between ICSI children and those conceived by traditional IVF. Because of the uncertainty as to whether there is an increased risk of chromosome abnormalities, most centers performing ICSI recommend chromosome testing of the fetus (by amniocentesis or chorionic villus screening). While we do know that apparently healthy babies can be born this way, the jury is still out regarding the future implications of this technique in terms of undetected genetic problems that may be present in these children.

"There's no more *halakhic* problem with ICSI than with any other method," declared Weitzman. But he went on to describe a dialogue between Rabbi Menachem Burstein, founder of Puah Institute, and an Israeli embryologist in the process of using ICSI to help an infertile couple conceive:

Rav Burstein once asked an embryologist, "Who decides which sperm to put in?" She answered, "I decide. I'm *Hakadosh barukh Hu* [The Holy One, blessed be He, or God], I decide."
"How do you decide?" he inquired.
"The one that moves the best, that's the one," she explained.
"*Hakadosh barukh Hu* makes *that* one move the best when you look in [the microscope], so *Hakadosh barukh Hu* decides," he countered.[22]

New evidence shows that when ICSI is performed to overcome male infertility, the percentage of embryos that reach the blastocyst stage of development (5–7 days after conception) is reduced by roughly 20 to 30 percent compared with straight IVF (using sperm that are capable of penetrating the egg on their own). Underlying problems with the sperm may cause this, since the chosen sperm, the ones that "move the best," could have abnormalities. It is also possible that the eggs are being killed by the process used to shove the sperm through the cell membrane. Thus, it is generally recommended that ICSI be used only in cases where the relative benefit of successful fertilization outweighs the potential risk—the loss of embryos.

Is Testicular Biopsy Permitted?

In order to perform ICSI, sometimes there are no sperm at all in the ejaculate, and it is necessary to aspirate, or draw out, immature sperm cells directly from the testis (this procedure is called testicular sperm extraction, or TESE).[23] Disruption of or causing harm to the testis is a clear prohibition in the Torah. Regarding humans, the Bible states, "A *petzua dakah* [man with crushed testicles] shall not enter into the assembly of the Lord" (Deut. 23:2), that is, he shall not marry an Israelite woman. Even castration of animals is expressly forbidden in the Bible. Referring to animal sacrifices, it is written, "One whose testicles are squeezed, crushed, torn or cut you shall not offer to God, nor shall you do these in your land" (Lev. 22:24).

Although the wounding or injuring of the testicles is prohibited, there may be instances when it is necessary to perform surgery on testes—in the case of testicular cancer, for instance, when the life of the man is at risk. In addition, in order to fulfill the mitzvah of *p'ru ur'vu* (be fruitful and multiply), it may be permissible to aspirate cells or even biopsy (cut out a section of) the testis in order to obtain immature sperm. Pieces of tissue from the testicle can be used to obtain sperm. And even if only a few sperm cells are isolated from the testis, those cells can be mechanically injected, one by one, into individual eggs. "Rabbi [Moshe] Feinstein . . . sanctions testicular biopsy," reports Rabbi Yoel Jakobovits, referring to the twentieth-century rabbinic scholar. "The Talmudic constraints are applicable only when the perforation of the testis results in infertility; nowadays the procedure has the reverse likelihood, being designed to help alleviate infertility."[24]

Does Judaism Allow Artificial Insemination?

Of course if there are no sperm whatsoever produced in the testis, then even ICSI cannot help. In that case the couple may consider using donor sperm. The procedure is referred to as donor insemination (DI) or artificial insemination—donor (AID).

The reactions of the broad spectrum within Judaism to this issue can provide insights into how different denominations consider other questions as well. We have seen in many bioethical issues that there is more that unites Jews from across the religious spectrum

than separates them. However, there are some dramatic differences in
the approach of more stringent vs. more liberal Jews to certain questions. With regard to the use of sperm donors, Orthodox sources are
concerned with the letter of the law—how halakhic precedent deals
with the issues. Conservative and more liberal sources address not
only halakhic issues, but social issues as well.

Rabbi Elliot Dorff, presenting "officially held positions of the
Conservative movement," declares, "The couple is . . . using DI
when they have no other way to achieve a precious goal in Jewish
law and thought, the bearing of children. Even if the social father
does not technically fulfill the obligation to procreate through DI,
we should applaud the couple's willingness to use DI for three reasons: the Jewish tradition has always valued children; in their efforts
to have children, couples who use DI will undergo hardships that
other couples need not endure . . . and they consequently need every
encouragement they can get; and, finally, having and raising Jewish
children is a demographic imperative for the Jewish community in
our time."[25]

Dorff's discussion of DI also focuses on the impact of the procedure on the marriage. For instance, should the child be informed of
his beginnings, or is secrecy appropriate? Is there asymmetry in the
family relationships; that is to say, does Dad treat the child differently
because he is not the genetic father? Dorff also considers demographic
concerns, issues of racism (the perceived need to use Jewish sperm),
and the issue of compassion, which is so vital to the welfare of an infertile couple coping with difficult options.[26]

Orthodox sources analyze the issue and dissect the pros and cons
from a halakhic and traditional point of view. For instance, the question arises as to whether there is precedence for conception without
sexual intercourse. The possibility of pregnancy without sexual intercourse was considered in the Talmud. *Hagigah* 15a recounts the possibility of a woman conceiving in a public bath into which a male has
discharged semen. If a woman maintains that she is a virgin, then this
possibility is considered. And she may still be eligible to marry a high
priest on account of having possibly conceived *sine concubito* (without intercourse).[27]

Artificial insemination has been performed for many years, and the
question of the halakhic validity of this procedure has been discussed
by many sources. It is clear that more rabbinic authorities approve of

artificial insemination if the husband's sperm is used (as long as it is not wasted in the process). However, the idea of using donor sperm has been controversial.

"Are there children in the *haredi* community who are born from egg donors?" asked Rabbi Gideon Weitzman, referring to the ultra-Orthodox population in Israel. "The answer is yes, and not just one [child]. There are children in the *haredi* community born from sperm donors, under *halakhic* guidance." He explained that, as long as the sperm donor is not Jewish, "there are certain cases where that is the option that is given the *heter* [rabbinic approval]."[28] Weitzman thus revealed a little discussed fact—that in the most observant Jewish communities in the world there are women conceiving with sperm from non-Jewish donors, with the permission of and under the guidance of their rabbinic leaders.

In one of the most controversial rulings regarding infertility, Rabbi Moshe Feinstein wrote a responsum in 1959 permitting the use of donor semen, as long as a non-Jewish donor was used.[29] With a Jewish donor, it is possible that, in the future, the offspring might establish an incestuous relationship. When the donor is not Jewish, he has no halakhic familial relationship to his offspring or to any other member of the Jewish community. Since the mother is Jewish, the child will be Jewish regardless of the origin of the sperm. Subsequent responsa written by Feinstein also support this view.[30]

Fred Rosner, on the other hand, presents some of the conflicting rabbinic opinions and troublesome issues involved. "Artificial insemination using the semen of a donor other than the husband (A.I.D.) is considered by most rabbis to be strictly prohibited for a variety of reasons, including the possibility of incest among the children, confused genealogy and problems related to inheritance. However, without a sexual act involved, the woman is not guilty of adultery and is not prohibited from living with her husband. The child born from A.I.D. does not carry any stigma of illegitimacy."[31]

"While it is not as acceptable as the husband's own sperm, a non-Jew's sperm is nevertheless preferable to an 'outside' Jewish male's contribution," explains Jakobovits. "The child born to a Jewish mother and a non-Jewish father is fully Jewish. A child born of a Jewish woman and Jewish father other than the lawful husband is a *mamzer*—a bastard—which in Jewish law has important consequential marital disabilities."[32]

Paternity with Donor Sperm: Who Is the Father?

The status of the halakhic father determines the status of his child with regard to the three groups, or castes, in the Jewish community, the Kohanim (Priests), the Levites, and the Israelites. This status was important in ancient times for assignment to different roles in Temple service; in modern times membership in one of the three groups determines certain rights and obligations in synagogue service. Rabbinic sources generally agree that paternity is determined by who provides the sperm, so that a baby conceived from donor sperm would not, halakhically, be considered the child of the infertile husband. Orthodox rabbi Kenneth Brander writes, "The male donor is considered the father of the child. This determines the child's status as a Jew: as a Kohen, Levi or Yisrael. It determines for which people the child is required to observe the laws of mourning. It defines which relatives the child is forbidden to marry; and it is instrumental in deciding issues of inheritance."[33]

Rosner's conclusions regarding the status of the child survey a wide range of rabbinic opinions.

Most consider the offspring to be legitimate. . . . [A] small minority of rabbis consider the child illegitimate; and at least two authorities take a middle view and label the child a *sofek mamzer* [of questionable legitimacy]. Considerable rabbinic opinion regards the child (legitimate or illegitimate) to be the son of the donor in all respects (i.e., inheritance, support, custody, incest, levirate marriage, and the like). Some regard the child to be the donor's son only in some respects but not others. Some rabbis state that although the child is considered the donor's son in all respects, the donor has not fulfilled the commandment of procreation. A minority of rabbinic authorities asserts that the child is not considered the donor's son at all.[34]

The range of views appears to be broad and to encompass all possibilities. However, even within the most observant communities there is some acceptance, and the use of donor sperm—as revealed above—does occur. AID is sanctioned and advised by some of the strictest authorities.

As an infertile couple in the 1980s, my husband and I were privy to advice—mostly unsolicited—from many camps. We had our share of relatives who admonished us for "waiting." We also had advice from *frum* [observant] people, most of whom had no clue what our specific problem was. One very religious relative let us in on that ultimate secret of the religious community—that donor insemination is permitted. (We did not opt for donor insemination, however; in fact, we conceived our daughters the natural way.) My reaction to this at the time was to consider that donor insemination may be very common in the Orthodox community. The actual figures on percentage of couples who resort to DI are not, to my knowledge, available.

The irony of using donor sperm from a non-Jew is that although a Gentile is not halakhically related, the possibility of genetic (and legal) incest—although unlikely—still exists! According to Dorff, in the case of a non-Jewish donor, "sexual intercourse between the people born through his sperm donation and those born through his marriage would not technically constitute a violation of Judaism's laws prohibiting incest." In fact, he adds, there is no halakhic incest "even if the non-Jewish donor's wife is Jewish and thus his children are Jewish, for Jewish law does not recognize family lineage among non-Jews through the father's line."[35]

Dorff reported an interesting difference in the practical application of donor insemination within the Jewish population. "Rabbi Feinstein's original position . . . has led to a curious result. Physicians report that while traditional Jews who use DI prefer non-Jewish donors for fear of incest in the next generation, liberal Jews want Jewish donors. The motivations for that tendency may be many, but undoubtedly for some people insemination by a non-Jew smacks of intermarriage, and others probably hold an ethnic notion of Jewish identity. . . ." Thus, he concludes, some Jews—including both Orthodox and liberal Jews—do not want to "pollute purity of the Jewish genetic line."[36] (Although the Orthodox and the liberal Jews are concerned about different forms of "pollution.")

Dorff expresses his concern on the social outcomes resulting from donor insemination. For instance, he worries that the child may feel awkward in synagogue if the social father is a Kohen or Levite, and he—being the child of a Gentile man—is designated differently.[37] According to Rabbi Shlomo Auerbach, when a non-Jewish donor is used, the child inherits the status of the mother. If she is a *bat Kohen* [daughter of a Kohen] or *bat Levi* [daughter of a Levite] then the child inherits that designation—from his maternal grandfather.[38]

Dorff also considers an extreme bizarre possibility. Since the social father has no halakhic relationship (and no genetic relationship either) to the child, sexual relations between social father and daughter are technically permissible. "Jewish marital law, though, must recognize the strong bonds that social parents create between themselves and the children they raise and among the children themselves. . . . Consequently sexual relations between the parents and children or between the children themselves are prohibited in the second degree."[39]

The use of donor sperm and the use of ICSI open doors to even more startling possibilities. If you can inject a sperm into a human egg and initiate embryonic development, why not inject a nucleus from a somatic cell (containing all forty-six chromosomes) into an empty egg? In that case, there is only one parent, the nucleus donor, and that procedure is called cloning. This is no longer a science fiction scenario; it may be accomplished in humans in the near future. And we will need to provide answers to complex questions in Jewish law and practice when the day comes that the first Jewish baby is cloned (see chapter 5).

CHAPTER 4

Embryonic Stem Cells:
When Does Life Begin?

ONE OF THE most significant biological developments of our time is the discovery of embryonic stem cells. These amazing cells, derived from human embryos, are capable of developing into just about any part of the body. The possible applications of stem cells for medical science are staggering. Stem cells may provide a way to grow tissues and organs to replace diseased body parts; and understanding stem cell biology may lead to treatments or even cures for cancer. However, since stem cells are generally obtained from human embryos and fetuses, there may be serious ethical problems regarding their use, and theologians and politicians are grappling with these issues.

Although sperm and eggs are living cells, they clearly do not have the same status as a fully formed human; and although fertilized eggs are alive, and have the potential to become fully formed humans, they also do not carry the same status—in legal, halakhic, or ethical terms—as a fully formed human.

When I first held a test-tube baby in my hand nineteen years ago I realized the potential of that technique, not only to help infertile couples, but also to eventually improve on nature. I was always compulsively careful with each petri dish and its precious microscopic conceptus. Who knew the future of any given egg or embryo? Each fertilized egg had the potential to become a human. Was I holding the next Albert Einstein or Marie Pasteur? But the potential cannot be realized unless the embryo is implanted into a woman's uterus.

"You can discard it, you can do medical research on it, you can freeze it for implantation in the future," Rabbi Tzvi Flaum said, referring to the human preembryo, or fertilized egg, produced by in vitro fertilization. Rabbi Flaum presented this startling rabbinic view at a workshop on infertility at an annual convention of the Association of Orthodox Jewish Scientists. Rabbi Flaum continued, "The preembryo is only a potential human being. . . . There is a broad consensus of the *poskim* [authorities on Jewish law] regarding the preembryo. At the insertion of the zygote [fertilized egg] into the uterus, issues of abortion begin."[1]

Rabbi J. David Bleich discusses Judaism's view of the "sub-visual," that which cannot be seen with the naked eye. According to Bleich, Jewish law does not concern itself with phenomena that can only be viewed using a microscope or magnifying glass.[2] For instance, although creeping animals are not considered kosher, microorganisms found in water are not prohibited, since they cannot be seen with the naked eye and, for all intents and purposes, do not exist according to Jewish law. Therefore, the early stages of a preembryo, which are still microscopic, may be of no consequence in terms of Jewish law—and may be destroyed at will. On the other hand, many rabbis emphasize that the subvisual can be consequential, thus a strong argument could be made regarding the importance and significance of the microsopic embryo.

"Halakha has no objection to using such an early stage of an embryo," writes Rabbi Moshe Tendler, a noted authority on medical ethics. "The halakha is quite clear. Such research should be encouraged. A fertilized egg in a petri dish does not have 'humanhood.' Without implantation into a uterus it remains a 'zygote' or pre-embryo and is not viewed as an 'abortus' as the church views it."[3]

Tendler refers to embryonic stem cell studies as "Remarkable life-saving research! . . . No one doubts the potential success of this effort. In other words, no one sees how it can go wrong. They should only be right and G-d should help guide their hands."[4]

The Fetal Tissue Resolution of the Union of American Hebrew Congregations states, "There is an emerging consensus of Reform Jewish authorities that tissue obtained from either therapeutic or spontaneous abortions may be used for purposes of life-saving or life-enhancing research and treatment. Jewish requirements that we use our God-given knowledge to heal people, together with the concept of *pikuakh nefesh* (the primary responsibility to save human life, which overrides almost all other laws) has been used by Jewish legal authorities to justify a broad range of organ transplants and medical experimentation. These requirements likewise justify the use of fetal tissue transplants."[5]

Thus, although embryonic stem cell research has caused a flurry of controversy in many camps, rabbinic sources appear to accept, even embrace, these developments.

How Are Stem Cells Produced?

When a sperm and an egg meet they produce a zygote, a single cell, which is the fertilized egg. That cell has the potential to produce a new human being. After the cell cleaves into two cells, both cells carry genetic instructions to make a whole human. Occasionally, in fact, the two cells disconnect from each other and they can develop into two separate, complete individuals, so-called monozygotic or identical twins. That can occur because both of the cells are still totipotent; they both harbor all the instructions and the capability to develop into a complete human being, including all of its approximately two hundred different cell types and parts, such as heart, lungs, liver, brain, skin, and blood.

In the typical course of human development, the fertilized egg divides into an embryo with two cells, four, eight, sixteen, thirty-two cells, and so on. Then that tiny microscopic cluster of dozens of cells begins to change its form. It becomes a hollow ball of cells; and within that ball, attached to the inner wall, a handful of cells cluster together forming the "inner cell mass." In the late 1990s it was discovered that cells from the inner cell mass of a human embryo could be collected and grown in culture dishes. (The embryo is destroyed in the process of harvesting such cells.) And those cells, under carefully controlled conditions, could be coaxed into forming nervous tissue, heart muscle, pancreatic cells, and blood cells, to name a few. So it was clear that special cells from an embryo, called stem cells, retain the ability to form different parts of the body. Stem cells can be either totipotent, which means they can form *any* other cell type, or pluripotent, which means they can form a *limited menu* of cell types.

In November 1998, a team of American and Israeli scientists first reported the growth of human embryonic stem cells in culture—that is, in a petri dish nourished by special nutrient medium.[6] In that study, Dr. James Thomson of University of Wisconsin at Madison joined forces with Dr. Joseph Itskovitz-Eldor, an in vitro fertilization (IVF) specialist from Technion, the Israel Institute of Technology in Haifa. Itskovitz-Eldor provided human embryos obtained from infertile couples participating in IVF. The embryos used were "surplus," as they were left over from couples who produced more embryos than necessary and had chosen to donate the extra ones for medical research.

Around the same time, a different team of scientists from Johns Hopkins University also developed a method to isolate stem cells. They found that primordial germ cells (PGCs) from aborted human fetal tissue also remained totipotent.[7] PGCs are cells in the fetus that eventually become sperm and eggs; they appear to maintain the ability to develop into many different cell types.

Reactions of Governments and Religious Leaders

Legislation during the Clinton administration banned federal funding of laboratory work to obtain embryonic cells from human embryos, but permitted federal funding of research on human embryonic cells once they were obtained. Thus, many researchers who required such cells collaborated with labs funded by biotechnology companies and other private sources. The private funds paid for the first step of the process, that is, taking cells out of embryos. In many cases federal grants supported the subsequent research on the resulting cell cultures.

Renewed debate on the issue of human embryo research led to the 1999 report of the U.S. National Bioethics Advisory Commission, which stated, "In our view the ban [on embryo research] conflicts with several of the ethical goals of medicine, especially healing, prevention and research."[8] In addition the report stipulated that "Research that involves the destruction of embryos remaining after infertility treatments is permissible when there is good reason to believe that this destruction is necessary to develop cures for life-threatening or severely debilitating diseases."[9]

On December 19, 2000, the British Parliament voted overwhelmingly (366 to 174) to allow greater freedom in embryonic stem-cell research. The legislators stated that, "Research using human embryos

... to increase understanding about human disease and disorders and their cell-based treatments should be permitted," subject to controls laid down in existing legislation.[10] This British legislation rekindled the controversy among political and religious leaders in Europe and the United States.

The Catholic view on the sanctity of life has influenced many politicians in the United States and abroad. As expressed by Reverend J. D. Cassidy: "The mystery of each new creation at human conception, the belief that the human soul is created by a direct act of God at the time of fertilization and that each person is unique and irreplaceable, shapes the Roman Church's doctrine that the human embryo is a human being. Fertilization represents the scientifically and morally logical place to draw the line between what is and is not human life."[11]

Not surprisingly, the Vatican has come out strongly against embryonic stem cell research. Pope John Paul II told scientists at an international conference in August 2000 that human stem cell cloning is "morally unacceptable." The Pope stated, "Every medical procedure performed on the human person is subject to limits: not just the limits of what is technically possible, but also limits determined by respect for human nature itself." He continued, "What is technically possible, is not for that reason alone morally admissible."[12]

Not all Catholics oppose embryonic stem cell research. Australian priest and philosopher Norman Ford asked, "Are all cells derived from an embryo, themselves embryos?" He asserts that stem cells taken from the inner cell mass cannot develop into a self-sufficient embryo, so should be permissible to use in research.[13]

Although the Vatican, which is located in Rome, is strongly opposed to stem cell research, Italian scientists have encouraged research into use of stem cells to address a range of degenerative diseases. On December 28, 2000, just days after the British Parliamentary vote, a group of twenty-five experts, led by Nobel laureate Renato Dulbecco, presented Italy's health minister with a report. The report of the Italian panel supports stem cell research; however, it recommends the use of stem cells derived from adult tissues, rather than embryonic stem cells, whenever possible. (Note that it does not recommend a ban on embryonic stem cell research.)[14]

Could adult stem cells be used in place of embryonic stem cells for the production and growth of cells and tissues? Although most cells of the adult become committed to a particular cell lineage, it is believed that some cells retain the ability to differentiate into a limited set of tissues, thus they would be considered "pluripotent" (as opposed to totipotent embryonic cells, which can develop into just about any cell type). Recent studies by several U.S. research teams showed that stem cells derived from adult tissue may be able to develop into a smorgasbord of cell types, just as embryonic stem cells do. A Princeton study, for instance, demonstrated that adult mouse brain cells can become liver, heart, muscle, and other tissue types. Use of adult stem cells could reduce and perhaps obviate the need for embryonic stem cells, which is clearly abhorrent to the Roman Catholic Church.[15] Unfortunately, most researchers in the field are not convinced that adult stem cells will be as versatile or as useful as their embryonic counterparts.

Discussion in Germany on this issue has been heated. The *Embryonenschutzgesetz* (law on embryo protection) is extremely restrictive;[16] and German biologist Gisel Badura-Lotter proposed at the Annecy Conference that current restrictions on embryo research be continued in Germany until it is clear that animal models and adult stem cells are inadequate to move research ahead. "Human embryos and fetuses have at least some sort of moral status that protects them from an uncontrolled use for the purpose of others," stated Badura-Lotter.[17]

The European Group on Ethics in Science and Technology (which advises the European Commission) voiced its concern that embryos would be created for the express purpose of producing stem cells. In a report published November 14, 2000, this group argued that "the creation of embryos with gametes donated for the purpose of stem-cell procurement [is] unacceptable, when spare embryos represent a ready alternative source."[18]

In the United States, the proposed guidelines for human stem cell research developed by the National Institutes of Health (NIH) suggested a cautious acceptance of some applications of the technology. The "NIH understands and respects the ethical, legal and social issues relevant to human pluripotent stem cell research and is sensitive to the need to subject it to oversight more stringent than associated with the traditional NIH scientific peer review process. In light of these issues, the NIH plans to move forward in a careful and deliberate way prior to funding any research utilizing human pluripotent stem cells."[19]

The NIH draft guidelines proposed the establishment of a Human Pluripotent Stem Cell Review Group to oversee compliance with the guidelines in funded projects. Under those guidelines, areas of research that would not be eligible for NIH funding would include projects involving the production of human clones; the combination of human stem cells with animal embryos; nuclear transfer into eggs to produce stem cells; and production of embryos expressly for the purpose of making stem cells.[20]

NIH guidelines stipulated that only surplus embryos—created but no longer needed for IVF—be used. They specified that strict protocols for informed consent of embryo donors must be followed. The guidelines prohibited payment of donors, and donors were prohibited from receiving any commercial benefits. Federal funding could not be used to actually obtain the cells. But once the cells are produced, further research on them may be funded.[21] (As described above, this type of policy has already led to interesting collaborative arrangements where private companies fund the procurement of stem cells in one lab, then those cells are transferred to a second lab where federal funds support the remainder of the project.)

Even politicians who are traditionally "pro-life" do not universally reject support for stem cell research. For instance Connie Mack, a former Florida senator and a pro-life Catholic, indicated his support for stem cell research. "For me, as long as that fertilized egg is not destined to be placed in a uterus, it cannot become life." Senator Orrin Hatch, who has been staunchly anti-abortion, came out in support of human embryonic stem cell research. In fact, a *Newsweek* poll reported that 57 percent of abortion opponents and 72 percent of Roman Catholics support embryonic stem cell research.[22]

On August 9, 2001, President George W. Bush revealed his proposed policy on stem cell research. His policy banned federal support of any further procurement of stem cells, and restricted federal support of research only to stem cell lines developed before August 9, 2001. Bush stated that there were more than sixty stem cell lines already in existence as of that date, and he asserted that those cells, growing in a handful of labs in the United States and abroad, should adequately provide material for stem cell research and applications. The stem cells lines that had been established by that cut-off date were maintained in laboratories in Singapore, Australia, Israel, the Netherlands, Norfolk, Virginia, and Madison, Wisconsin.[23]

Bush's decision was heralded by some as a compromise, but denounced by others (on both ends of the political spectrum) as a sell-out.

Some right-to-life factions were upset that human embryonic cells would be used in research at all. On the other hand, many scientists voiced concern that the policy would be too restrictive. There was concern that there actually were many fewer usable cell lines than first reported and that existing cell lines were controlled by private concerns (hence access may be restricted).[24] In addition, most of the lines were generated in the presence of mouse cells. Mouse cells provide an environment that enhances the growth and survival of human stem cells. However, human stem cells grown in this way may be not be usable for human therapy, as they may be exposed to mouse viruses—and there is concern that mouse viruses could cross species and infect human cells.[25]

Jewish Views Permit Stem Cell Research

During the vigorous discussion and debate in spring and summer 2001, many Jewish groups voiced their views on the issue. Both Hadassah: The Women's Zionist Organization of America, and the National Council of Jewish Women indicated support for stem cell research. Rabbi Elliot Dorff is quoted as indicating the use of embryos for research is a "mitzvah." "It's not only permitted, there is a Jewish mandate to do so." He also indicated that creating an embryo for production of stem cells is permissible, but "less morally justifiable." Reform Judaism's Union of American Hebrew Congregations (UAHC) sent letters to the Bush administration in support of "carefully regulated" federal funding of the research.[26]

A joint letter from the Union of Orthodox Jewish Congregations of America and the Rabbinical Council of America expressed their "support for federal funding for embryonic stem cell research to be conducted under carefully crafted and well-monitored guidelines." The letter continued, "Our Torah tradition places great value upon human life; we are taught in the opening chapters of Genesis that each human was created in G-d's very image. The potential to save and heal human lives is an integral part of valuing human life from the traditional Jewish perspective. Moreover, our rabbinic authorities inform us that an isolated fertilized egg does not enjoy the full status of person-hood and its attendant protections. Thus, if embryonic stem cell research can help us preserve and heal humans with greater success, and does not require or encourage the destruction of life in the process, it ought to be pursued."[27]

As indicated above, Rabbis appear to accept human embryonic research—even though Judaism has strong views on the sanctity of human life. Jewish principles of *pikuakh nefesh* (preservation of life) and *v'rappo yirappe* (and heal, he shall heal, Exod. 21:19) supersede many restrictions and commandments. *Pikuakh nefesh* even overrides the observance of the Sabbath. Thus, if embryonic stem cell research has therapeutic applications and is used to save a person's life, then it can be justified even though human embryos are destroyed in the process.

Talmudic references to the early embryo guide modern Rabbinic decisions; the Talmud refers to a human embryo of less than forty days as *maya be'alma,* or mere water. According to Rabbi Yitzchok Breitowitz, since a fetus that is less than forty days in development is considered by the Talmud as "mere water," a woman who suffers a miscarriage in the first forty days does not acquire the status of *tumat leidah*—the impure state following the birth of a baby. Breitowitz, Associate Professor of Law, University of Maryland, cites Rabbi Mordechai Eliyahu (the former Sephardic Chief Rabbi of Israel), Rabbi Chaim David Halevi (the Ashkenazic Chief Rabbi of Tel Aviv), and Rabbi Moshe Sternbuch to support the view that "Most contemporary *poskim* . . . have allowed the destruction or at least the passive discarding of 'unwanted' preembryos, ruling that the strictures against abortion apply only to embryos or fetuses within a woman's womb and not to preembryos existing outside of it."[28]

"Since the preembryo is not in an environment in which it will be able to be brought to term and live, it arguably does not have the status of a living being, even according to those who might accord such status to an implanted embryo before 40 days," maintains Breitowitz. "It has also been suggested that regardless of the forty-day rule, until there is implantation within a human being, no human life can be said to exist. (It might also be added that since the preembryo is microscopic, not visible to the unaided human eye, the Torah does not invest its existence with any halachic significance.)"[29] (Also see above regarding the status of the subvisual in Jewish law.)

Breitowitz does voice concern that destruction of a preembryo may violate the prohibition of *hashkhatat hazera* [improper emission of seed].[30] However, many rabbinic authorities agree that *hashkhatat hazera* only applies to semen per se and is not violated after an egg is fertilized.

What about the concern over *ba'al tash-khit* (do not destroy)? Could destruction of an embryo for the purposes of producing therapeutic cells be wanton destructfulness? Maimonides discusses the issue of *ba'al tash-khit*, declaring: "It is not only forbidden to destroy fruit-bearing trees but whoever breaks vessels, tears clothes, demolishes a building, stops up a fountain or wastes food, in a destructive way, offends against the law of 'thou shalt not destroy.'"[31] However, when you destroy something in the process of producing something useful, it does not appear to violate *ba'al tash-khit*. As the popular saying goes, "you can't make an omelet without breaking some eggs." Similarly it would seem that destruction of an embryo to produce a therapeutic treatment is not wanton wastefulness.

"Is there a mitzvah to preserve preembryos?" asks Breitowitz. The commandment *vahai bahem* (and you shall live by them, Lev. 18:5), for instance, would suggest that we are commanded to protect even a preembryonic life. On the contrary, Breitowitz argues. Rabbinic sources agree that one may not violate Shabbat on behalf of a preembryo. He concludes that "there are no affirmative obligations to sustain preembryo life, so the surplus embryos may indeed be discarded with impunity for any reason."[32]

In his book *Be Fruitful and Multiply,* Dr. Richard Grazi explains, "Nontransplanted embryos fertilized artificially *in vitro* have no standing as fetuses in Jewish law. Rabbi [Chaim David] Halevi rules that 'all eggs fertilized *in vitro* have no standing as embryos . . . and one may discard them if they were not chosen for implantation, as the law of abortion applies only to [procedures in] the womb. . . . But *in vitro*, as was said, there is no prohibition at all.'" Grazi also quotes the ruling of Rabbi Mordechai Eliyahu who wrote, "those eggs which have not been chosen for implantation may be discarded."[33]

Grazi also cites the views of Rabbi J. David Bleich, an expert in Jewish bioethics, who has pointed out that an aborted fetus in the early stages of gestation has a different status and does not require burial. "Rabbi Bleich has noted that the tissue of a spontaneously aborted fetus may be used for research purposes if there is a reasonable basis for assuming that practical medical benefits will ensue within a reasonable time period," Grazi reports. "One might well argue that untransplanted embryos, which never had viable status, might be used for the purposes of research before they are discarded."[34]

The Central Conference of American Rabbis, a Reform Jewish association, has summarized Jewish thought on the issue of preembryo disposal in a succinct and clear manner. Although the CCAR holds very liberal views on abortion and the right to choose, they also saw fit to base their analysis on the Talmudic *maya be'alma* (mere water), and rulings of Halevi and Eliyahu. The CCAR responsum concludes, "A human embryo or zygote is, like the fetus, a potential but not a legal person, and there is no explicit Jewish legal prohibition against its destruction. . . . The embryo may be used for medical research, provided that it is handled with the proper respect and reverence."[35]

The use of stem cells to treat diseases will quickly come to pass, as PPL Therapeutics, the company who produced Dolly the cloned sheep, is initiating work on the use of stem cells to create insulin-producing pancreas cells for the treatment of diabetes. Additional diseases to be targeted by other groups in the near future include Parkinson's disease, Alzheimer's disease, and spinal cord injuries. Researchers at Memorial Sloan-Kettering Cancer Center in New York turned mouse stem cells into dopamine-producing brain cells, which one day could be used to cure Parkinson's disease. Other studies at New York's Rockefeller University transformed mouse stem cells into pancreatic islet cells—a potential cure for juvenile diabetes.[36] In addition, two research studies at the Technion-Israel Institute of Technology and Rambam Hospital in Haifa reported on the transformation of embryonic stem cells into heart tissue to replace scarred heart muscle and pancreas cells to treat diabetes.[37]

And Rabbi Moshe Tendler forecasts that the study of stem cells will have an impact on our understanding of cancer. "It will also teach us why some cells refuse to differentiate and become cancer cells," declares Tendler. "If I know how cells differentiate, I may then find out why some cells do *not* differentiate. Therefore we look upon stem cell research as the only open avenue for cancer research today."[38]

Meanwhile, experimenters from the Jones Institute, a preeminent fertility clinic and research center in Norfolk, Virginia, reported that they actually had created embryos for the express purpose of extracting stem cells. The *New York Times* report referred to this accomplishment as "breaking a taboo against creating human embryos expressly for medical experiments. . . ." The newspaper report states, "What some see as promising research others see as ghoulish," but concludes that "in the future there may be reason to create human embryos to derive stem cells—not for research, but for treatment, so that a person suffering from a disease could be treated with cells that are an exact genetic match. Such embryos would be created using cloning technology, an issue that is itself hugely controversial."[39] (See chapter 5.)

Technology is already able to support the beginning and the end of human gestation; as technology advances it may well lead to an artificial womb to be used for all nine months of gestation. One day, when a woman's uterus becomes superfluous, the status of the human preembryo and subsequent stages of development will have to be reconsidered.

According to Rabbi Richard Address, Director of UAHC Department of Jewish Family Concerns (bioethics program of UAHC), "What will be an issue, is technology—as it pushes the frontier of what is possible in utero. . . . The beginning of life issue is fascinating and may be the next true frontier to be dealt with."[40]

CHAPTER 5

Bone of My Bones and
Flesh of My Flesh: Human Cloning

There will come a time when science will know how to create human beings without the natural intimate act. This has been explained in the books of science and is not an impossibility. —COMMENTARY OF THE MEIRI[1]

And the Lord God caused a deep sleep to fall upon the man, and he slept; and He took one of his ribs, and closed up the place with flesh instead thereof. And the rib, which the Lord God had taken from the man, made He a woman, and brought her unto the man. And the man said: "This is now bone of my bones, and flesh of my flesh; she shall be called Woman, because she was taken out of Man."

—GEN. 2:21–23

THIS STORY OF Eve's creation is the first report of adult cells being used to create an entirely new individual. Although the creation of Eve was not precisely cloning (the production of a genetically identical individual), the miracle described in Genesis foreshadows new scientific developments and applications that were never previously imagined possible. The breakthrough in animal cloning occurred in 1997 in a laboratory in Scotland where scientists produced a lamb that is a clone—that is, a genetically identical copy—of an adult sheep.

The first laboratory cloning of a mammal was accomplished by taking cells from an adult sheep and fusing them with unfertilized sheep eggs (whose nucleus containing its genes had previously been removed). The genes from the adult cells began to direct the activity of the eggs, and the eggs developed into embryos. The cloned embryos were implanted into surrogate mother sheep. Only one of the cloned embryos survived and developed to term. The lamb that was born was an exact genetic copy, or clone, of the original donor sheep. That cloned lamb was genetically identical to the original sheep, just as identical twins are genetically identical to each other.[2]

Scientists were excited about this development since it was previously believed that cloning a copy of an adult animal would be impossible. When animals develop from embryo to adult, their cells specialize, and the genes of different organs and tissues become committed to certain tasks. For instance, liver cells specialize to perform liver functions and brain cells specialize to perform brain functions. In specialized cells, genes for other functions become inactive, and, it was thought, the inactive genes are "turned off" forever. Cloning experiments performed on frogs and toads in the 1950s through 1970s yielded limited success. It seemed that only embryonic cells could be reprogrammed to direct eggs to develop; and none of the cloned frogs or toads lived to adulthood.

The successful cloning experiment performed in Scotland proved that the genes could be "turned on" again, to direct all of the developmental steps to produce a whole, functioning adult animal. The breakthrough involved the production of mammalian clones; thus, scientists anticipated the new technology being applied to producing genetic clones of valuable livestock. For instance, livestock that are disease resistant, cows that are exceptional milk producers, and animals that have been genetically engineered to make pharmaceutical drugs could all be cloned. Some of those experiments have already been successfully accomplished.

The ability to clone a sheep presents us with the specter of human cloning. In the 1970s scientists who were concerned with new developments in reproductive biology and biotechnology discussed and debated the implications of human cloning. During that period the novel, and later the popular movie, *The Boys From Brazil* presented the scenario of Hitler-clones being raised by unsuspecting adoptive families all over the world.[3] *In His Image: The Cloning of a Man*[4] was a supposedly nonfiction book that claimed human cloning had secretly been performed using cells from an eccentric multimillionaire who wanted immortality. That story turned out to be a hoax.

Should the cloning of humans be limited or prohibited? Justifications for human cloning could include altruistic ones, such as attaining better understanding of human development and/or disease, or enriching the world by producing another "copy" of a brilliant scientist, humanitarian, talented musician, artist, or athlete. In various articles written in response to the cloning of Dolly the sheep, individuals who have been "nominated" for cloning include Michael Jordan, David Letterman, Hillary Clinton, Donald Trump, Dolly Parton, and Albert Einstein. My biology class nominated Elvis Presley. Unfortunately, it is too late for Einstein or Elvis, since dead men can't be cloned (unless their cells are carefully preserved at the time of death). With regard to the other nominees, I am sure that the world can survive very well with just a single copy of each of them (although it might be an even more interesting and entertaining world if there were more Steven Speilbergs). Motives for cloning could easily be less than noble, such as producing clones for spare body parts for one's self, or to attain genetic immortality.

If you know any sets of identical twins (nature's clones), or are a twin yourself, you will appreciate that, even though they may look alike and share many other similar traits, identical twins are unique individuals with regard to personality, and may have many other significant mental and physical differences. The differences between identical twins occur as a result of environmental influences, such as differences in nutrition, communicable diseases, accidents, emotional factors, and educational opportunity. There are many factors that shape personality, and the mechanism by which they influence development are not well understood. Just as identical twins differ, artificially generated clones who share the same genes will also differ considerably as the result of environmental influences. So if you are considering having yourself cloned, be aware that your clone most definitely won't be another you. In fact, it will be an infant at first, so, unless you can convince your own mother to do it again, you'd better be prepared for a difficult job of child rearing.

The successful cloning of Dolly the sheep in 1997 opened up a Pandora's box of speculations and concerns in the Jewish world. Many of the concerns regarding cloning involve whether humans should be cloned, and what would be the repercussions of such activities. The motives and goals of human cloning have since been vigorously discussed.

Legislation on Human Cloning

In the discussion on embryonic stem cell research (also see chapter 4), it has been noted that cloned human embryos prepared from a patient's own cells could be used therapeutically (rather than grown into a baby). The cloned embryos would provide stem cells genetically identical to that patient, and those stem cells could be used to grow tissues or organs that genetically match the patient. This type of tissue would not be rejected by the patient's immune system. (Immune rejection is a common problem when transplanted tissue from other sources is used.)

Despite promising therapeutic applications, many countries have outlawed human cloning. In 2001 the U.S. House of Representatives voted to ban all human cloning (as of this writing the Senate has yet to act). Current United States legislation does, however, block the use of federal funding for cloning, and without federal money most research into human cloning will be difficult to accomplish. There have been reports of private groups who are intent on producing a human clone in the near future. For instance, an article in the *New York Times Magazine* reports that a "science-loving, alien-fixated sect called the Raelians" are actively pursuing the cloning of a human being. They call their cloning venture Clonaid. A prime candidate for cloning is a ten-month-old infant who died accidentally and whose parents had cells frozen for that purpose. The article reports that "The Raelians have at least 50 female followers volunteering as egg donors and surrogate mothers in the Clonaid project."[5]

The Israeli Knesset has passed a law prohibiting human cloning. Rabbi Dr. Avraham Steinberg, head of the Medical Ethics Program at Hebrew University in Jerusalem participated in the Knesset committee that developed the law. He reported, "Initially the law was phrased in a way that human cloning, and research in human cloning is forbidden forever. To me it sounded like a very bad law." Steinberg explained that, in his view, cloning is probably halakhically permitted and may contribute positively to human health.[6]

"Moreover, I claimed that even if the Knesset will ban [cloning], if it will be invented in England or Japan, it will come anyway," Steinberg continued. "So who are we, the small Knesset of small Israel to do such an overriding prohibition that forever any research involved should be outlawed?" Steinberg's influence helped change the formulation of the bill to one that bans human cloning for a period of five years, but permits research on the activation of cells and production of human embryonic tissues, "without actually getting to a human clone."[7]

The most significant legislation to date was passed in January 2001, when the British House of Lords approved a proposed law (which had already been passed by the House of Commons in December 2000) permitting the creation of cloned human embryos. The British have been at the forefront of human embryo research since the birth of the first test-tube baby, Louise Brown, in England in 1978. Although there was a flurry of protests regarding in vitro fertilization when that breakthrough was announced more than two decades ago, British scientists and subsequently researchers from the United States, Australia, and dozens of other countries, persisted in their work in order to learn more about the early steps of development of the human embryo. The British scientific establishment was initially more liberal in allowing research on human embryos. They pioneered work that enabled scientists to improve the technique of IVF—leading, ultimately to much higher rates of pregnancies and births for infertile couples. Some of the work was ethically questionable; for instance, some experiments involved growing embryos for weeks in petri dishes, beyond the point where an embryo should implant into the uterus.

The British legislation permits cloning of embryos for production of stem cells (see chapter 4), which, essentially, is only one step removed from the cloning of a human baby. By the time you read this chapter, it is possible that human cloning will already be a fait accompli. And after a flurry of protests and a ripple through society, public interest in human cloning may quiet down; scientists will knuckle down to determine how to apply this technology for improving human health (and making lots of money).

The Science of Cloning

Since February 1997, when Ian Wilmut and his colleagues in Scotland first announced the cloning of Dolly the sheep, ethicists, theologians, scientists, and legislators the world over have been grappling with the issues involved. In the original experiment, a cloned sheep was produced by removing the nucleus—the compartment containing most of the genes—from a sheep's egg and replacing it with genetic material from an adult sheep cell. This led to the creation of a genetic duplicate of that adult sheep.[8] Although this procedure has been replicated in cattle, mice, and other species, it is still far from being a routine procedure.

In order to accomplish human cloning, human eggs would be obtained from a donor female's ovaries. In the laboratory, the nucleus of each egg, containing the genes of the egg donor, would be removed and replaced with a nucleus from the person to be cloned (the clonor). Most adult cells carry nuclei that could, theoretically, be used to create clones since those cell structures hold all of the genes necessary to produce a human being. After the egg has been engineered to hold the genes of the clonor, it becomes an embryo, a potential human being. The embryo is placed into the uterus of a woman who carries the pregnancy as the gestational and birth mother.

The cloned child, or clonee, born from such a procedure would be genetically identical to the clonor—the donor of genetic material. That child would not be genetically related to the gestational mother, unless the gestational mother herself provided the genetic material. Although the clonee would not be genetically identical to the egg donor, it should be noted that the donated egg—sans nucleus—would still have a small but significant genetic contribution to the child. In addition to the nucleus, every egg carries mitochondria, cell structures outside of the nucleus, that contain about 1 percent of the genes of the cells. Thus, the mitochondrial DNA of the egg donor (about 1 percent of its total DNA) would be inherited by the clonee.

Is Cloning Kosher?

While no clear consensus exists as to whether human cloning is "kosher," Jewish scholars have analyzed the situation and have identified some major issues, concerns, and possible ramifications of cloning vis-à-vis the Jewish faith and the Jewish community. Several of the leading Jewish medical ethicists have addressed the difficult question posed by Rabbi Michael Broyde: "Is cloning an intrinsically good, bad or neutral activity?"[9]

A wide spectrum of reactions has unfolded regarding the new technology. In fact, when scientists first speculated that mammals could be cloned, early rabbinic reactions included that of Rabbi Azriel Rosen-feld, who worried that cloning could destroy family relationships. He suggested that the methodology could result in a person being born without a halakhic father. Nevertheless, Rosenfeld concluded that cloning can be permitted because this reproductive method does not involve a sex act—thus, it is not halakhically forbidden.[10]

Dr. Fred Rosner's views on cloning initially focused on the princi-ple that halakha is said to apply only to what is visible with the naked eye. Rosner objected, however, to giving a blanket approval of clon-ing, explaining, "There are three partners in the creation of a human being: The mother, the father and God. Cloning of man negates iden-tifiable parenthood and would thus seem objectionable to Judaism."[11]

The Conservative view is accepting of human cloning, despite fears voiced by those who imagine armies of exact duplicates. "Religion has nothing to fear from science," insists Rabbi David Wolpe, assist-ant to the Chancellor and Professor of Modern Jewish Thought at the Jewish Theological Seminary. "You can't make an exact replica of a human being because a human being has something unfathomable—no matter what technology is used. The great divide is between mate-rialists, who think physical matter is everything, and those who think human beings contain something intangible."[12]

Dr Stephen Modell, writing in Reform Judaism's *Study Guide XI: Infertility and Reproduction,* points out that it is important to con-sider risks involved with human cloning, a process that is still poorly understood by scientists. "Bodily or 'somatic cells' are known to col-lect mutations at a certain frequency throughout life. Normally these natural mutations disappear when the cell dies off, but here they are being allowed to remain and pass into subsequent generations." Mod-ell emphasizes, "If cloning were to be applied to human beings at the current stage, the level of embryo discard would be unacceptable by either secular or religious standards."[13]

Derekh Hateva: *According to the Natural Order*

One critical issue, according to Jewish scholars, is whether cloning vi-olates the dictum of *derekh hateva* (according to the natural order). Dr. John Loike of Columbia University and Rabbi Dr. Avraham Stein-berg published concerns on this point:

The idea of creating human life via replication . . . may go against a general dogma that God directed human life to be formed through natural sexual processes. . . . First, no *zera* [seed] from the man is used to create the child, and second, the creation of human life can be achieved in the absence of men or male-derived tissue. . . . The *gemara* . . . states [Nidda 31a and Kidushin 30b] that three partners (God, man and woman) are required for the creation of a human being, and that *zera* of both man and woman contribute to the development of the child.[14]

Steinberg points out that, while the Jewish view is to be very cautious in accepting new things, there is a general principle by the commentary *Tiferet Yisrael* that could apply to new situations: "Anything which we have no reason to prohibit is permitted, without having to find a reason for its permissibility. For the Torah does not mention every permissible thing, but rather only those things which are forbidden."[15]

Cloning may also be permitted, perhaps even encouraged by Jewish law, because of the passage *asher boroh Elokim laasot* (which God created to produce, Gen. 2:3). The term "to produce" should have been in the past tense, *v'asah* (and he produced). "To produce" refers to man's role in continuing the process. "Hashem, by his will did not complete the world," explains Steinberg. "He left it for human beings to complete the world. . . . We are permitted to interfere in nature. . . . [W]e are obligated to interfere, obligated to improve the world."[16]

The biblical command *umilu et haaretz vikivshuha* (fill the earth and conquer it, Gen. 1:28) also indicates that "we are partners with [God] . . . to improve the world. It is not an option—it is an obligation to continue to improve the world and do good for the world," affirms Steinberg. Steinberg explains that as long as the act of perfecting the world does not violate halakhic prohibitions, or lead to results that would be halakhically prohibited, then we are given a mandate to use science and technology to improve the world. For instance, if a mule is produced by intermixing two species, this may be an improvement of certain qualities of the animals that are useful to man. However the act of intermingling two species is expressly prohibited by the Torah. So Jews are not allowed to interfere in nature in this way. Similarly, if cloning results in production of offspring who are judged to be illegitimate *(mamzerim)*, then it would be forbidden.[17]

Ironically, cloning may not simply be kosher, it may actually be a preferred method of reproduction in some instances. The problems with many reproductive technologies involve the procurement and use of sperm. There is concern by many rabbinic authorities that artificial insemination as well as in vitro fertilization would violate the prohibition of the "wasting of seed."[18] In addition, use of a donor sperm for a married woman carries the risk of technical adultery and might result in her offspring being of questionable lineage. Cloning would obviate those problems because semen is not used at all in the process.

Is a Clone a Golem?

A significant concern regarding cloning is the issue of humans attaining the ability to create life. Steinberg explains that true creation is production of something from nothing *(yesh me'ain)*, life from nonlife, and only God can accomplish that. The process of cloning, on the other hand, involves producing life from preexisting life, something from something *(yesh me'yesh)*. Therefore we are not trying to usurp God's powers of creation.[19]

Interestingly enough, the Talmud has recorded instances of rabbinic leaders producing living creatures—like the legendary golem, an artificial anthropoid—from dust.[20] In talmudic accounts a golem was formed from earth and was activated by putting the name of God either on the golem's forehead or on a manuscript inserted into his mouth. A golem was incapable of speech, and since speech was said to be associated with possession of a soul it was concluded that a golem did not possess a soul. This would explain why renowned rabbinic leaders were reportedly able to destroy a golem—without violating the prohibition of murder.[21]

One question that rabbis have addressed is whether a clone is considered a human being or is more analogous to a golem. Bleich discusses the differences between a golem and a clone. He explains that for a golem, the "replication of already existing human genetic material is completely lacking."[22]

Clones are clearly different from golems in the way they are generated, as well. Clones are produced by inserting the nuclear DNA from an adult cell into an egg whose nuclear DNA has been removed. The egg is gestated in a woman's uterus, and the clone is born in the usual way. "A clone, no less that any other 'born' child, meets the prima-facie test for humanness and is human," declares Broyde.[23]

Rabbi J. David Bleich concurs. "The crucial distinction between a golem and a clone is that a golem . . . clearly lacks a human progenitor. A human clone, although the product of asexual reproduction, does have a human progenitor."[24]

In fact, in defining characteristics of human beings, Bleich adds another revelation. "Offspring produced from a cloned cell of a monkey or chimpanzee implanted in a human womb," he asserts, "although having both the genotype and phenotype of an animal, would be regarded as human for the purposes of Jewish law."[25] He cites Talmud *Niddah* 23b, which indicates that "an animal-like creature born of a human mother is regarded as a human being."

The definition of humanness, being born of a human mother, should suffice given the current state of cloning technology, which still requires a human female to carry the pregnancy. However, what if, one day, gestation could occur completely outside of the womb?

"The Talmudic conclusion seems to be simple," Broyde explains. "When dealing with a 'creature' that does not conform to the simple definition of humanness—born from a human mother—one examines context to determine if it is human. Does it study Torah (differential equations would do fine for this purpose too) or is it at the pulling end of a plow? By that measure, a clone, even one fully incubated artificially, would be human, as it would have human intellectual ability and human attributes."[26]

Does a Clone Have a Soul?

An interesting Reform responsum, from the Central Conference of American Rabbis, addresses the question of whether a clone possesses a soul. This document first deals with the question of when a soul enters the human body. The responsum explains, "There is considerable discussion among the rabbis about the moment at which the soul enters the body. Is it at the instant of conception, of embryonic formation, or of birth? All three were possible for those rabbis who followed the Neo-Platonic three-fold division of the soul." Accordingly, at each of those stages, one element of the soul would enter the body. Maimonides and other Jewish philosophers of the Middle Ages designated the three forms of the soul as animal, vegetative, and human, indicating their belief that even lower forms of living organisms possessed souls.[27]

"Anything produced asexually like a clone would be akin to a plant, would also be considered to have at least a lower form of the soul. The soul in its human form, according to *halakhic* tradition, however, enters a *body* only at the time of birth," maintains the responsum. As we've described above, it will be necessary, at least initially, for a cloned human to be gestated and born from a human mother; thus presumably it would acquire a human soul. At some point in the future when it is possible for human embryonic development to occur independently of a mother's womb, this issue will be even more complex. The responsum concludes, however, that "we could well consider such a being to have a soul. It will have been formed from human material despite all genetic alterations. Its development will have taken place in an artificial environment rather than the womb, but at some point it will emerge as a human being. . . ." The responsum ends on this curious note: "Unless such possibilities of independent intellectual and moral development are genetically removed, this would be a human being."[28]

As we draw closer to identifying genetic elements involved in moral development, it may be possible to modify humans and alter their capacity for moral judgment. (Note that researchers reported in 2001 that specific regions of the brain involved in moral reasoning have been identified.[29]) If the intellectual abilities and moral development are compromised, would that then permit us to define such a being as something less than human? And where do we draw the line?

What Are the Family Relationships of a Clonee?

Cloning can be performed using the nuclear DNA from a man's or a woman's cell. As described above, the nuclear DNA from the donor (clonor) is inserted into an egg from which the nuclear DNA has been removed. Nuclear DNA makes up about 99 percent of a person's genes. The other 1 percent are located in cell structures called mitochondria, which reside in the cytoplasm, outside of the cell's nucleus.

When a man's DNA is used, the clonee produced in this way will have a combination of mostly the man's (clonor's) genes plus a small portion of a female egg donor's genes. In addition, the reprogrammed egg needs to be incubated in a woman's uterus. Thus, there are at least two, and potentially three participants in this process (if the egg donor and the woman who provides the womb are two separate individuals). A cloned boy produced in this way will be almost identical genetically to the man who was cloned. There will be a small amount of genetic similarity to the egg donor due to the 1 percent of genes from the egg's mitochondrial DNA. The woman who provides only a uterus will not be genetically related to the clonee at all.

It is interesting to note that children have already been born carrying the genes of two mothers. They were generated by a variation of in vitro fertilization where a small amount of cytoplasm—including mitochondria—gets transferred from a donor egg to a recipient egg. This is done in order to increase the fertilizability of the recipient egg. The children conceived from those eggs retain a small portion of mitochondrial genetic material from the woman who provided egg cytoplasm.

When a woman's DNA is used, it is inserted into an egg as described above, and also gestated in a woman's uterus. However, in this case, there could be one, two, or three participants, since the same woman, or two or three different women, could theoretically provide the DNA, the egg, and the womb. No male participation whatsoever is needed for this process.

In Jewish law, it is often critical to determine parentage as this impacts on family obligations, obligations in community service and the critical question of "Who is a Jew?" Although Reform Judaism now recognizes paternal as well as maternal Jewish descent, traditionally, the Jewish status of a child is transmitted through the mother. Given the complexity of the relationships between clonor and clonee, and the other parties involved, "Who is the mother?" and "Who is the father?" may become difficult questions to answer. Two scenarios will now be discussed:

1. When DNA from a man is used for cloning, producing a male clonee:

WHO IS THE MOTHER?
Since an egg (relieved of its own nuclear DNA) is used, and a woman's uterus is needed, then the egg donor and/or the gestational woman could be considered the mother. If two women were involved—one as egg donor, and the other gestational, it is possible that both would be afforded the status of mother. According to Bleich's

analysis of the halakhic implications of surrogate motherhood, it is possible for a child to have two halakhic mothers.[30]

WHO IS THE FATHER?

If the father is only defined as the one who donates sperm, then it is possible that the male DNA donor may not be considered the father because no sperm is involved.

If so, there is no father. Alternatively, the male DNA donor would certainly be considered the father because he gives twice the amount of DNA as a male does in a normal conception. According to Broyde, "A man who reproduces through In vitro fertilization contributes only half of the genetic material through his sperm, and is still considered the father according to normative Jewish law. . . . Certainly in this case, the fact that the man contributed all of the nucleic genetic material would appear to be enough to label this person the father according to Jewish law. . . ."[31]

2. When DNA from a woman is used for cloning, producing a female clonee:

WHO IS THE MOTHER?

Would it be the female DNA donor, the egg donor, or the woman who carries the pregnancy? Based on rabbinic analysis of in vitro fertilization, egg donation, and surrogate motherhood, the woman who gives birth is clearly accepted as a mother. As described above, Bleich suggests, however, that the egg donor has at least an equal claim on this status, and thus there may be two halakhic mothers. According to views of the late Rabbi Shlomo Zalman Auerbach, as cited by Bleich, if both egg donor and surrogate are considered mothers, and if either the egg donor or the gestational mother is not Jewish, the Jewish status of the child would be in doubt, and the child would have to be converted to be considered a Jew.[32] Thus, when a female is cloned, the situation could entail one, two, or three halakhic mothers: the DNA donor, the egg donor, and the gestational woman.

WHO IS THE FATHER?

Since there is no semen involved, perhaps the female is both mother and father. However, this contradicts prevailing views. "When establishing the identity of the mother and the father, Jewish law insists that only men can be the father and only women can be the mother," declares Broyde.[33] One exception in the Talmud, involving an individual who is a hermaphrodite, suggests that fatherhood and motherhood could be difficult to define and ambiguous under very unusual circumstances. Broyde cites a Talmudic account of an androgynous male "who fathers a male child and then has a (homo)sexual relationship with that male child. . . ." The rabbis question whether the transgression is only incest, or both incest and homosexuality. If the androgynous male is not considered a true male, then having sexual relations with his son would only violate the prohibition against incest. If he is considered a true male, then he would be violating prohibitions against incest and homosexuality. This shows that a hermaphrodite might be defined as a mother and/or a father, thus the role of father is not necessarily restricted to a male.[34]

Alternatively, the father of the female clonor could be considered the clonee's father. The glaring problem with this scenario is, if the female clonor is the mother, then her father should be the grandfather, not the father. "To rule that . . . the father of the provider of the genetic material is the father—seems far removed from logic," argues Broyde, "as that person is completely uninvolved in the reproductive process. The one who fertilized the egg, either by providing half the normal chromosomes in the case of regular fertilization, or all the chromosomes, in the case of cloning, should be the parent."[35]

Finally, it is possible that there is no halakhic father. Loike and Steinberg suggest that a fatherless female clonee may be analogous to a *shetuki*—a person who does not know who his or her father is. The sages declared that a *shetuki* is not permitted to marry anyone, because he/she might marry a half sibling and produce an offspring from an incestuous relationship, a *mamzer*. A female clonee, however, cannot be a *shetuki* as there is no issue of her not identifying a father, because, in fact, she has no father at all. One other precedent for a person not having a father occurs when someone converts to Judaism. After conversion all genetic ties are halakhically broken, and the convert is considered to have no blood relatives from before conversion *(gerut)*. Thus a convert has no halakhic father or mother. This status may most closely approximate the female clone's fatherlessness in terms of her definition in the Jewish community. "*Gerut,* however, represents a complicated situation, and rabbinical authorities must determine if it is a valid precedent to our cloning situation to establish whether a cloned child lacks a halakhic father," conclude Loike and Steinberg.[36]

3. Who Are the Siblings of a Clonee?

Since a clonor provides an exact copy of his/her genes to produce a clonee, the relationship is similar to identical twin siblings. Therefore, should the clonee be considered a sibling of the clonor? "The definition of siblings found in Jewish law is either a common mother or a common father or both," explains Broyde. "[In most cases] clonor and clonee do not share a mother (egg donor or gene provider) or a father (provider of genetic material) and thus are not siblings. . . ."[37]

The one clear exception to this would be if parents clone their own child, using the DNA of the original child and the egg of the mother. In that case, the clonor and clonee would share the same mother, and would be, for all intents and purposes, like identical twins born years apart.

Does Producing a Clone Fulfill the Biblical Obligation "Be Fruitful and Multiply"?

Every Jewish male has an obligation to reproduce, based on the biblical injunction "be fruitful and multiply" (Gen. 1:28) (women are exempt) and the lesser rabbinic obligation to "inhabit the earth" (Isa. 45:18). Does cloning fulfill this obligation?

In the production of a female clone, no male is involved. Thus, production of a female clonee clearly does not fulfill any male's obligation to reproduce. In the production of a male clone, it depends on how reproduction is defined. Bleich's view is that a male is required to contribute semen to fulfill the biblical requirement. His argument is based on the phrase "male and female did He create them" (Gen. 1:27). "It would thus seem that, even if actual cohabitation is not required," Bleich reasons, "the commandment is nevertheless fulfilled only if the child is the product of gametes contributed by both the male and the female."[38]

Broyde is inclined to credit a male clonor with fulfillment of a mitzvah. Although some rabbis have ruled that sexual relations are required to establish paternity, most rabbis have accepted children produced through assisted reproductive methods as legal children of the inseminator. Broyde reasons (as quoted above) that the contribution of the father in IVF is only 50 percent of the genes, and he is judged to be the father. In the production of a male clone almost 100 percent of the genes come from the cloner. Thus, he should certainly be labeled the father, and be given credit for the mitzvah of *pru ur'vu* (be fruitful and multiply).[39]

Do the Risks Outweigh the Benefits?

Steinberg emphasizes that, for any new technology, we need to determine whether the good outweighs the bad. For instance, the automobile and airplane are both positive technological developments. But both also cause death and injury. In Israel, in fact, automobile fatalities exceed all other causes of death. Cars and planes are not banned, because the positive contributions to society and the world far exceed the drawbacks.[40] On the basis of risk/benefit, should we accept cloning technologies or reject them?

Some major benefits may include applications leading to the treatment of human disease. The principle of *v'rappo yirappe* (and heal, he shall heal) would apply. Cloning could be used in repair or replacement of damaged or dead cells, treatment of cancer, production of organs or bone marrow for transplantation, and addressing the needs of infertile couples. For instance, human cloning might present a way to repair or replace damaged brain cells in patients with Parkinson's and Alzheimer's disease; cloned embryos could be used to generate stem cells for such applications (chapter 4). Cloning could lead to production of animals, or even people, to be used as donors (organ or bone marrow) to treat deadly diseases such as leukemia. And cloning could provide a way for an infertile couple to produce a child who is genetically related to at least one of the parents.

Rabbis have considered the possibility of producing a human clone to act as a donor for an ill child. "Jewish law and ethics see nothing wrong with having children for a multiplicity of motives other than one's desire to 'be fruitful and multiply' (Gen. 1:28)," indicates Rabbi Michael Broyde. "There is no reason to assert that one who has a child because this child will save the life of another is doing anything other than two good deeds—having a child and saving the life of another. . . . Such conduct should be encouraged rather than discouraged."[41]

Bleich also cited medical applications as justification for human cloning. "There have been unfortunate cases of children afflicted with leukemia whose only chance of survival is a bone marrow transplant," Bleich explains. "If cloning were available, parents, in such rare situations, could clone the ill child. The newly-born infant would be disease-free but would be genetically identical to its afflicted sibling. Medically, the child would be an ideal donor."[42] Of course if the leukemia arose as the result of inherited genetic factors, the clone might also develop the disease.

The drawbacks to human cloning include the possibility that clones would not be treated as humans, and would be abused in some way. Cloning might be used eugenically, to create people with special characteristics: a tall child, or a smart child, a child who is musically inclined, etc. Some people are disturbed by the potential abuse of the power to shop for specific traits. Other abuses could include production of large groups of identical people to perform certain tasks, as portrayed in Aldous Huxley's *Brave New World*.[43] In addition, cloning may change the social order in ways that are difficult to predict. In many cases it would be hard to define who the father or mother is; therefore, some are concerned that cloning could lead to the destruction of the nuclear family.

The status of human beings is a major concern expressed by Dorff, who is worried that human clones may be treated like commodities. "If cloning is left to the economic forces of the marketplace, presumably it is the rich and the famous, but not necessarily the good, who would be cloned." And if standards are determined as to who should be cloned, "those most useful to society today will not necessarily be the most useful to society tomorrow." Dorff further argues, "whereas the labor market commodifies people's skills, cloning commodifies their very being. This certainly denies the inherently sacred character of human life affirmed by the Jewish tradition."[44]

So, despite the potential life-saving applications of cloning, rabbis continue to stress caution. Bleich does voice concern regarding the applications society may develop vis-à-vis this technology. Nevertheless, he continues to support the notion that if cloning can hold out hope for new therapeutic approaches, "despite the attendant risks, [cloning] may be regarded as moral and even laudatory. . . ."[45]

Ecological Ramifications

The ability to produce genetically identical herds of sheep, cattle, and flocks of millions of cloned chickens and other farm animals will affect agriculture. Farmers are already striving for genetic homogeneity within their animal populations. It is easier to mass produce meat and other animal products if the quality is predictable and certain traits are present throughout the animal population. For instance, dairy farmers would prefer to cultivate herds of cows that are all excellent milk producers. Using animal husbandry, which involves the cross-breeding of animals with the most desired traits, many farmers have already accomplished some of these goals. However, when sexual reproduction is used (in classical animal husbandry, for instance) there are still no guarantees regarding traits of individual animals, and genetic diversity is still present in the population.

On the other hand, cloning produces an exact genetic copy of an animal. Dorff cautions that animal cloning should be used judiciously because of the issue of genetic diversity. "If the only species of sheep bred in the future is the one that produces the most wool, all other species of sheep will eventually die out. What happens, then, if a particular virus or bacterium attacks that specific variety of sheep? The whole sheep population could disappear."[46]

When diseases strike diverse populations, even if some animals are susceptible, there are usually individuals able to survive and repopulate the species. That would not necessarily be the case in a population where every individual were exactly alike. "This problem illustrates just one aspect of the issue of *biodiversity,* the need to protect multiple forms of vegetable and animal life," states Dorff.[47]

Dorff also emphasizes that cloning may result in unpredictable outcomes. "The possibility—indeed, the likelihood—of mistakes in the cloning process makes cloning also a potential source of threats to the environment and to us humans," continues Dorff. "That is more than a general moral concern; it is a distinctly Jewish one. Adam's charge in the Garden of Eden was 'to work it and to preserve it' (Gen. 2:15), and ever since then ecological concerns have carried both theological and legal authority within Judaism. . . . If we do permit human cloning . . . we must take conscientious steps to prevent environmental catastrophe."[48]

This argument regarding the need for biodiversity might apply to human cloning as well, if large numbers of identical individuals were cloned, producing herds of identical humans. Loike and Steinberg raise the concern that a lack of genetic variation resulting from production of large groups of identical people would lead to a scenario where disease could wipe out large populations of people.[49] Diversity protects the human race from mass eradication by disease. Lack of diversity would expose the population to new risks.

Dr. Harvey Gordon, writing in the UAHC Program Guide X (1998) on "Cloning," asserts that "Be fruitful and multiply" (Gen. 1:28) refers to reproduction, not replication. The diversity in the human population has arisen from sexual reproduction. "Every descendant of Adam and Eve—from Cain and Abel, to you and me—has been a unique individual."[50]

Gordon writes that if God wanted us to copy ourselves, the commandment would read "*Replicate* yourselves." Cloning would thereby qualify to fulfill the dictates of "Fill the Earth." However, we are a diverse population and we do not reproduce naturally by replication, we do so by sexual reproduction—leading to new genetic combinations. "The diversity of our species suggests that what God has in mind is not replication, but *reproduction*," asserts Gordon. "Children are not genetic replicas of either parent. They are a *product* of their union; genetic heirs of both parents, they are genetically unique."[51]

The Talmud presents a compelling argument in favor of the diversity of humankind. "Man was created alone . . . to proclaim the greatness of the Holy One. For if a man strikes many coins from one die, they all resemble one another; in fact they are exactly alike. But although the king of kings, the Holy One, blessed be He, fashioned every man from the die of the first man, not a single one of them is like his fellow. Hence, each and every person should say, 'The world was created for my sake.'"[52]

Gordon concludes that human cloning does not address *pru ur'vu* as the mitzvah was intended by God. "It is nothing new for Reform Jews to look beneath the surface for the deeper meaning of *mitzvot*. . . . Each of us is created in God's image, yet each of us is unique. It's clear that *genetic diversity* is an important part of God's plan. That is the course set out upon by Adam and Eve. To that extent, human cloning does not respond to the first *mitzvah* given to our original ancestors, 'Be fruitful and multiply.'"[53]

Aldous Huxley's *Brave New World* left us with visions of armies of cloned people, engineered to do whatever job it is they have been designed for.[54] This specter can elicit the so-called "yuck factor"—a visceral disgust, experienced by some, to certain biotechnological applications. In this scenario, the armies of clones could produce new social hierarchies and might affect the ecology and balance of human populations within the environment. And Judaic thought—concerned with ecological balance and perturbation of the environment—is not so quick to accept such a scenario.

Speculations on the Future of Cloning

Rabbi Noam Marans, a Conservative rabbi, speaking at a symposium on bioethics, expressed concern that the process of cloning would yield offspring with abnormalities. When that occurs with animal cloning, the animals can be destroyed. However, Marans asked, "What are the risks in this area? . . . there's a possibility that there would be a lot of mistakes made. . . . What do we do with the mistakes? What if the mistakes are born? What if they have deformities? What if there are issues of abortion in this context? Are you allowed to experiment and abort as you go along, with a higher goal of reproducing—for people to have children?"[55]

The risks of cloning are just being realized. It took a full four years after the cloning of Dolly the sheep for scientists to admit that the technique is unpredictable and produces animals with random genetic errors. The success rate for cloning of most of the animal species studied is only about one percent. The other 99 percent of fetuses die during gestation or shortly after birth. In mice the success rate is slightly higher, about 3 percent. The main problem with cloning appears to be that, in order to build a complete animal, the DNA of the donor cell has to be reprogrammed by the egg's environment so that all the genes can once again become active. It appears that the reprogramming of genes that must occur for cloning to work can produce error-riddled DNA. Those genetic errors lead to the abnormal development and death of most of the clones.

The mammalian clones that survive have a wide range of defects, including "developmental delays, heart defects, lung problems, and malfunctioning immune systems." Genetic problems have been observed in almost every clone that has been produced. The few that survive and are born exhibit a very high rate of birth defects. In some animals the defects appear later in life. For instance, some cloned mice appeared quite normal through young adulthood but then suddenly developed extreme obesity—they grow enormously fat in middle age. Another common occurrence is "large calf syndrome." This was first observed in cows and describes the tendency of cloned animals to grow abnormally big, developing in large placentas, and being stillborn or dying soon after birth. Even Dolly, the first cloned sheep, who initially appeared normal, began to show signs of obesity as compared with normal sheep, and had to be put on a special diet.[56] She has also developed arthritis at a relatively young age.

Although rabbinic *poskim* (halakhic adjudicators) are still discussing the permissibility of human cloning, the fact that it is not yet perfected and could lead to serious defects will certainly be a factor in rabbinic approval. Because of these serious health issues, while cloning could one day prove to be useful therapeutically, it is premature to use it on human beings. "You can't engage in something of high health risk," cautioned Dr. John Loike. "The chance of success is very small. You are going to generate children who are deformed."[57]

"I cannot imagine actually doing this with humans until the science was at a point where we were fairly confident that kind of thing wouldn't happen," agreed Rabbi Moshe Shapiro. "I can't imagine how anybody with any ethical sensitivity would want to experiment in that way."[58]

"Jewish medical ethics is one area of Jewish law where we tend to have strong interdenominationalism, where Orthodox, Conservative, Reform, Reconstructionist, secular Jews, all tend to agree in general," concludes Rabbi Noam Marans. "On the issue of cloning, it seems to be that one cannot prohibit that which is not explicitly prohibited by Jewish law, but one must prevent those side effects, those risks that come out of something that is technically legal from the Jewish perspective. So what we have to do is say . . . 'Under what circumstances?' and how are we going to deal from a Jewish perspective with the side effects of cloning, such as 200, 300, 500, 700 botched experiments that may be human beings with problems?"[59]

On the one hand, according to Bleich, "tampering with natural processes in a manner that would lead to social upheaval is not included in man's mandate 'to fill the earth and conquer it' (Gen. 1:28)."[60]

However, the biblical commandment *v'rappo yirappe* (and heal, you shall heal, Exod. 21:19) appears to sanction, perhaps even require, humankind to use technology for positive medical applications. And cloning may eventually open new doors to novel medical treatments and new opportunities to understand disease and treat infertility.

Assuming that cloning can be developed to the point where the risks are low and the benefits are great, Jewish law may be able to sanction its use. "One is inclined to state that *halacha* probably views cloning as far less than the ideal way to reproduce people;" Broyde admits.[61]

"However," Broyde maintains, "when no other method is available it would appear that Jewish law accepts that having children through cloning is perhaps a *mitzvah* in a number of circumstances and is morally neutral in a number of other circumstances."[62]

In Genesis (2:23), the phrase "Bone of my bones and flesh of my flesh" describes the creation of Eve from Adam. God took cells from Adam's side and created Eve from those cells. Creating a female from a male is obviously not creation of an exact copy. However, it could theoretically be accomplished with some genetic manipulation, and could be considered a variation on the cloning theme. In nature, some insect females produce males from their own unfertilized eggs. In the Adam/Eve scenario a male could produce a cloned female by the following mechanism: When sperm cells are made by males, half of the sperm contain only the X chromosome and no Y (male) chromosome. Those X chromosome–containing sperm cells could be genetically manipulated (by doubling the chromosome number) to produce female cells. From those cells, a female of the same species could be created. In this way, Adam's cells could have been used to produce Eve, who would have "inherited" half of his genes, but would not have been an exact copy of him. She would have been "bone of his bones and flesh of his flesh," the first "cloned" human. Consider all the trouble *that* cloning experiment led to in the Garden of Eden.

CHAPTER 6

The Seven Deadly Diseases

I'M NOT A fan of needles. To be sure, the thought of giving blood makes me more than just a bit queasy. Yet my husband and I recently elected to have a blood test to determine if we are carriers of any of seven serious genetic disorders that are found in higher frequency in Ashkenazi (Eastern European) Jews. The opportunity to use new technology—DNA testing—to gain information about possible genetic problems was intriguing. It is also useful to know what disease genes run in the family, as it could have repercussions for our future grandchildren. And the tests, which can run upwards of $800 per couple and are not covered by most health insurance plans, were being offered free of charge thanks to a grant to the Saint Barnabas Hospital Jewish Genetic Disease Program.[1]

What would we hope to find out from these genetic tests? The tests are designed to determine if either of us is a carrier of any of seven deadly diseases. There are a handful of genetic diseases that are much more prevalent in people from Eastern European, or Ashkenazi Jewish, heritage than in the general population. The prevalence of genetic diseases is typical of a population that has been genetically isolated—or at least partially cloistered—for hundreds or thousands of years.

Jewish families carry specific genes at a higher rate than the general population for several reasons. For thousands of years, marriages would typically occur between individuals within the community; intermarriage was strongly discouraged. Conversion of non-Jews to Judaism was also discouraged. In addition, according to Dr. Robert

Desnick, the European Jewish community, estimated to number in the tens of thousands up through the fifteenth century, expanded in a population explosion during the subsequent four hundred years to the pre-Holocaust population of more than 6 million. Desnick, Chairman of the Human Genetics Department and Director of the Center for Jewish Genetic Diseases at Mount Sinai School of Medicine in New York, explained that the expansion from a relatively small number— called a "population bottleneck"—to a much greater number, resulted in a group of descendants who carry common genes from common ancestors.[2]

Genetic research has revealed that specific mutations—common altered genes inherited from common ancestors—appear at high frequencies in Ashkenazi Jews. It is thought that one common ancestor could be responsible for each of these novel, specific mutations in the DNA. This is called the "founder effect." For instance, there are two major mutations that cause Canavan disease in Ashkenazi Jewish individuals. They are called "E285A" and "Y231X."[3] They most likely arose from two separate individuals who served as "founders" of the mutation and passed those defective genes on to their unlucky descendants. Those descendants could now live throughout the Ashkenazi Jewish community worldwide. If two of those descendants who are carriers eventually get together and have children, then each child has a one in four chance of inheriting the disease.

In other words, a carrier of Canavan disease has one normal form of a gene, symbolized C, and one copy of the recessive disease gene, symbolized c. His genotype or complement for that gene is thus Cc. A spouse who is a carrier would likewise be designated as Cc. Each parent will transmit one form of the gene to each child. A child can thereby inherit C and C from both parents (25 percent chance), C from father and c from mother (25 percent), c from father and C from mother (25 percent) or c and c from both parents (25 percent). Thus there is a 25 percent chance for the child to be completely free of the Canavan gene (CC), a 50 percent chance for the child to be a carrier (Cc), and a 25 percent chance for the child to inherit Canavan disease (cc).

Humans are diploid; this means that we have a pair of each chromosome (numbered 1–22); we also carry a pair of sex chromosomes. Females have two X-chromosomes, and males carry one X- and one Y-chromosome (see chapter 8, on sex selection). Each set of twenty-three chromosomes comes from one parent—twenty-three from mother and twenty-three from father (total = 46 chromosomes). Scientists estimate that on the forty-six chromosomes there are about forty thousand genes that encode all the traits needed to make a human being. We carry two copies of most genes—one copy is inherited from each parent. Genes direct the production of proteins, which interact and thereby determine the structure and function of our cells and bodies. Each gene, therefore, affects the expression of a trait. Traits can include visible attributes, such as hair color, eye color, presence or absence of dimples or freckles, etc. Traits can also include biochemical differences, such as ability or inability to perform a chemical reaction crucial to the cell's survival.

When genetic errors, or mutations, occur, they can affect obvious traits. Many genes that control traits at the biochemical level can cause subtle changes, or they can wreak havoc with the body if they do not function properly. For instance, one gene that controls pigment formation can mutate and cause albinism, the complete lack of pigment. Albinos have white hair, very light skin, pink eyes (because with no pigment, blood vessels in the eye can be seen), and associated defects.

Some genetic diseases are dominant; this means that a person only has to inherit a single copy of the flawed gene to express the disease. Huntington's chorea is a lethal disease that is inherited through a dominant gene. A person only has to inherit one copy from one parent and he or she will develop Huntington's disease, a late onset, neurodegenerative disease that usually strikes in the fifties or older. Huntington's leads to progressive degeneration of function. Because the symptoms of the disease appear later in life, some people already have passed the gene on to their children before they are aware that they have the disease. Folksinger Woody Guthrie died of this disease in 1967. His widow helped organize support for research that eventually led to identification of the gene responsible for the disease.[4]

As explained above with regard to Canavan disease, many genetic diseases and conditions are carried by recessive forms of a gene. A person must inherit two copies of a recessive flawed gene, one from each parent, in order to have the disease. The cystic fibrosis (CF) gene is a recessive gene; thus, a child with CF has inherited a CF gene from each carrier parent. A carrier does not have symptoms, because one functional copy of the gene overrides the effects of the flawed copy. It is only when both members of a couple are carriers of the *same* recessive genetic disorder that there is a 25 percent chance that a given child will inherit two copies of the disease genes and will have the genetic disease.

Most of the tests we elected to take at Saint Barnabas Hospital are based on DNA technology and were unavailable a few years ago. The only one that was available more than twenty-five years ago, when we were newlyweds, was a biochemical test to identify Tay-Sachs carriers. The original test for Tay-Sachs carrier status does not involve DNA technology at all; it involves measuring the activity of a protein—an enzyme—in the blood. The enzyme test is still considered a preferred mode of screening, since it picks up a wider range of genetic defects. By measuring the enzyme activity, the test is looking at the product of the gene; any serious defect in the gene will manifest itself in the way the enzyme works (or doesn't work). On the other hand, DNA testing zeroes in on specific mutations, or errors that typically appear in the gene—asking, for instance, "Is this error present?" "Is that error present?" Only the most common mutations are tested for; therefore, some of the less typical mutations could be overlooked and not picked up by DNA testing.

The severity of Tay-Sachs was also a strong motivation for early development of a screening test. Babies with Tay-Sachs appear healthy for the first few months of life, then begin to deteriorate, losing mental and physical abilities. As the disease progresses, a Tay-Sachs child becomes blind, deaf, unable to swallow, and eventually paralyzed, usually dying by ages three to five.

Back in the 1970s, before we were married, my husband participated in a Yeshiva University screening program to determine if he was a carrier of Tay-Sachs. His test came back negative; thus, later on it was not necessary for me to be tested. Since the gene that causes Tay-Sachs is recessive, both parents must be carriers for a Tay-Sachs baby to be conceived. Nowadays, screening centers encourage both members of a couple to be screened simultaneously. That way, the results for both partners are available at once. Thus, if one partner tests positive for a disease gene, it is not necessary to endure the tension of waiting a month or two to determine if the other partner is also a carrier. This is especially critical if the woman is already pregnant. In that case, both partners should certainly be tested at once, since any delay should be avoided.

The availability of DNA screening tests has ushered in a new diagnostic era, providing important information to prospective parents. These new developments are changing the way people ponder, plan, and live their lives.

According to Randi E. Zinberg, Director of the Genetic Counseling Program at Mount Sinai Medical Center in New York, in the Ashkenazi Jewish population, there is a one in twenty-five carrier frequency for Tay-Sachs disease, one in forty for Canavan disease, and one in seventy for Niemann-Pick disease (Type A). All three of these diseases are lethal within the first few years of life. Other serious diseases that are screened for include: Gaucher disease (Type 1), cystic fibrosis, Fanconi anemia (Type C), and Bloom syndrome.[5] (Since we were tested, an eighth disease, familial dysautonomia, has now been added to this panel of tests.)

"Of couples that screen for all seven diseases," reported Zinberg, "one in six individuals will be carriers for *one* of those diseases."[6] In order for an affected baby to be conceived, *both* parents must have copies of the same disease gene. Then, if they both do carry the same disease gene, there is a one in four risk of conceiving a child with that particular disease.

My husband and I tested negative for all seven diseases. That is a genetic legacy I am pleased to pass on to the next generation. And since we are negative for these mutations, our offspring are probably not carriers either. It is important to note here that these tests are not 100 percent certain, since they screen for the major Ashkenazi Jewish forms of each mutated gene, but not *all* possible variants. Accordingly, the letter that accompanies the laboratory report states: "In the Ashkenazi Jewish population the tests performed can identify 95 percent or more of carriers."[7]

How Do Screening Programs Work?

Mutations—or genetic errors—can occur in any cell of the body, including the sperm and eggs. Mutations that occur in a body cell can alter that cell in such a way that it becomes cancerous (see chapter 10). However, those genetic changes are not usually transmitted to the next generation. Only mutations that occur in sperm or eggs can remain in the germ line and be passed from generation to generation. Mutations that involve recessive genes can be passed from generation to generation silently, until two carriers get together and produce a child with two copies of the mutated gene.

"Although a carrier usually doesn't know he's a carrier, every one of us are carriers for many recessive disorders," explained Dr. Gideon Bach, head of the Department of Human Genetics in Jerusalem's Hadassah Hospital. "Today you don't have to give birth to an affected child . . . a blood test can determine if you are a carrier. And since many of these disorders are incurable, prevention is the most important tool."[8]

In fact, everyone is a carrier of some defective genes. Geneticists have estimated that each person carries, on average, five recessive mutations. Most of us are carriers of defective genes that are so rare that the chances of marrying someone with the same recessive gene may be one in a million, or even lower. If the carrier rate of a given defect is one in a thousand, then there is a one in a million chance ($1/1000 \times 1/1000$) for those people to get together randomly. That is why there are many very obscure genetic diseases that occur in only a handful of people in the United States. Of course human beings rarely mate randomly. People frequently choose mates preferentially from within certain groups. Many people marry within their own religious or ethnic group. In some non-Western societies choosing a mate from within the extended family (e.g., cousins) is preferred. This type of consanguineous marriage and other nonrandom matings can increase the incidence of diseases even when the carrier rate is very low.

Through screening programs, Ashkenazi Jewish couples can learn if both husband and wife are carriers of the same genetic disorder. Prenatal diagnosis can detect fetuses with lethal genetic disorders very early in a pregnancy. Screening programs such as the one at Hadassah Hospital have reduced the number of Israeli babies born with cystic fibrosis and Tay-Sachs to almost negligible numbers. "Today you see [patients with those diseases only] in textbooks . . . today you hardly see them at all. At least 90 percent have been eradicated or avoided," reported Bach.[9]

Even the ultra-Orthodox, or *haredi,* Jews in Israel participate in screening for genetic diseases. However, since *haredi* rabbis usually do not permit abortions, in that community couples are prescreened before the young men and women are even introduced to each other. In the *haredi* community, most couples meet via matchmakers; a program called Dor Yeshorim seeks to ensure that men and women who are brought together as couples do not carry genes for the same genetic disorder.

"It is a screening project for various Ashkenazi disorders, but it is pre-engagement screening," Bach explained. "Almost all engagements are prearranged today [in that community]. The couple will not meet until they have a genetic I.D. It is done anonymously, and they don't get the results unless both partners are carriers of the same disease." In those cases, the "match" is simply never made. "They participate in this program very enthusiastically," said Bach. "At least 99 percent of their youth is tested."[10]

Dr. Robert Desnick helped to establish the Dor Yeshorim program in the United States to provide screening services for the ultra-Orthodox community. "In the Hasidic community, prenatal diagnosis is not feasible, abortion is not permitted, artificial insemination with donor sperm and birth control are not options," reasoned Desnick. When a match is proposed by the matchmaker, Desnick reported, "both parties call the central office to find out if the match is compatible." By keeping all medical information confidential, neither party is stigmatized as a carrier of a particular genetic disorder. When a couple is determined to be incompatible for a genetic reason, according to Desnick, "the engagement is avoided." Since this system was developed in 1983, more than 230 matches of Tay-Sachs carriers have been prevented. That program and other screening programs have led to an unexpected phenomenon—that Tay-Sachs associations in the United States are now predominantly filled with, and run by, non-Jewish couples.[11]

The National Tay-Sachs and Allied Diseases Association (NTSADA) has developed a program to encourage genetic screening of Reform and Conservative couples who are about to be married. This group has sent letters to hundreds of rabbis, asking them to "join us in reaching out to couples who are about to marry and start a family."[12]

"By advising all brides and grooms to have their Tay-Sachs and Canavan Diseases Carrier Tests, couples, on the threshold of their lives together, can have a means to safeguard their future families against the tragedy of Tay-Sachs, Canavan and the allied diseases. . . . Unfortunately, babies have recently been born in our area, with these diseases," states the letter signed by Marion Yanovsky, Co-President of NTSADA New York Area, and Lawrence Shapiro, Director of Medical Genetics at New York Medical College.[13]

NTSADA has prepared "Gift of Life" cards, which are meant to be distributed by rabbis to couples they are about to marry. They inform the couple, "We know you're busy, but we ask you to add just one more thing to your checklist. . . . Arrange for your carrier test today. It takes a few minutes to protect your future and your family. . . ."[14]

Yanovsky, speaking at the L'Dor V'Dor, From Generation to Generation, Health Education Conference, noted that as a result of the success of screening programs, "many rabbis, doctors and people in general think that the disease has been cured, or doesn't exist anymore. And people often do not think to test. . . . If you don't test for your carrier status, the only other way to find out if you're a carrier is to, God forbid, have a child with the disease," Yanovsky cautioned. "Therefore, we have to constantly pursue the path of carrier testing, because there is no cure [now] and [any future cure] is quite a distance away."[15]

"Recently there was a young couple in Rockland County [in New York state] with a Tay-Sachs child," Yanovsky continued. "Their rabbi never told them to be tested. And they had relatives who were tested who never told them [to be tested]." Since rabbis commonly counsel couples before a marriage is performed, the NTSADA established the program, in cooperation with the New York Board of Rabbis, designed to educate rabbis with regard to genetic testing. Yanovsky expressed frustration at the rabbis' low level of response; only a small percentage of them appeared to follow up on the information. She encouraged rabbis to pass on the information by distributing the Gift of Life cards to couples, and by counseling couples to be tested.[16]

Is There Fallibility of Screening?

Despite campaigns to screen Jewish couples, some babies have been born with those dreaded genetic diseases. The reasons for this are complex. First of all, ironically, it is the success of the program that leads to some of these births. Since Tay-Sachs has declined so dramatically in the Jewish community, some might assume that it is no longer a threat. Recessive genes do not just disappear from a population; in actuality, the rate of Tay-Sachs carriers in the population is no different than it was thirty years ago. The difference is that screening has reduced marriage rates between carriers, and when two carriers are married, fetal screening helps couples to avoid Tay-Sachs births.

Another factor involved has to do with the accuracy of testing. The Mount Sinai School of Medicine Genetic Testing Laboratory report qualifies its "negative" results for each test with the following disclaimer: "This negative result does not completely exclude the possibility that this patient is a carrier for . . . [name of disease] due to other mutations in the . . . [name of disease] gene."[17]

The report concludes with another important disclaimer: "This type of mutation analysis generally provides highly accurate genotype information. Despite this high level of accuracy, it should be kept in mind that there are many potential sources of diagnostic error, including misidentification of samples, polymorphisms, or other rare genetic variants that interfere with analysis. Families should understand that rare diagnostic errors may occur for these reasons."[18]

One dramatic development underscored the need for these types of disclaimers. "Tay-Sachs Alert," declared the ominous advertisement that appeared in American Jewish newspapers. The ad further explained, "If someone in your family was tested for the Tay-Sachs gene and the test was screened by Corning Clinical Laboratories, MetPath, MetWest or Quest Diagnostics between 1992 and 1998, please contact us to determine whether you need to schedule a free retest to confirm your results. For women and couples who are pregnant or plan to become pregnant, we urge you to contact us immediately." Unfortunately, this alert was too late for three couples who, despite being tested by these companies and informed that they were noncarriers, went on to have Tay-Sachs babies.

The most common test to determine Tay-Sachs carrier status relies on the ability to detect levels of the enzyme hexosaminidase A in the blood. A deficiency in that enzyme leads to a buildup of lipids in the nervous system resulting in irreversible damage to nerve cells. One in twenty-five Ashkenazi Jews is a carrier of Tay-Sachs and can transmit a single copy of the gene to their offspring. As described on page 88 for recessive disorders, when both parents are carriers, each child has a 25 percent chance of inheriting *two* copies of the gene, and thus getting the disease.

The three types of tests now available include a blood serum test for the enzyme, a leukocyte (white blood cell) test for the enzyme, and a DNA test. The Mount Sinai Center for Jewish Genetic Diseases advertises that its carrier screening for Tay-Sachs disease, which uses enzyme tests, has a "detection rate for carriers [of] approximately 99%."[19]

A commercial company involved in Tay-Sachs testing, Genzyme Genetics, reports a 98 percent sensitivity of their enzyme test, but only a 94 percent detection rate using DNA testing. This is because DNA tests are so specific that not all types of mutations are identified within the test. Thus, people with different mutations of that gene would go undetected. For that reason, Genzyme Genetics does not recommend using the DNA test alone, but rather in combination with, or as a confirmation of, the enzyme test results.[20]

"The assignment of Tay-Sachs carrier status is based on statistically derived cutoff levels and therefore is subject to some inherent limitations," Gary Samuels, spokesman for Quest Diagnostics, explained. He described how the test provides a range of values for enzyme activity. Patients who test at the upper end of the range are clearly noncarriers. Patients whose enzyme activity falls in the low end of the range are classified as carriers of the Tay-Sachs gene. For those whose range is in the middle, "there's an overlap between the upper end of the carrier range and the lower end of the noncarrier range and we call that gray zone the inconclusive range," reported Samuels. At the end of 1998 the company decided to interpret the test in a more conservative manner, since they had learned of Tay-Sachs babies born to couples whose tests were originally judged as negative. Samuels explained, "We expanded the inconclusive range to make the test more sensitive."[21]

As a result of this adjustment in ranges for enzyme activity, some patients originally designated as noncarriers now are considered to be "inconclusive." And Quest Diagnostics has therefore recommended retesting of those individuals. Of the 36,000 tests administered during 1992–98, Samuels estimated that 8,200 individuals would be recommended for retesting. Samuels emphasized, "We expect that 99% of the people recommended for retesting will continue to be [classified as] noncarriers."[22]

"In fact, we did this analysis in connection with a lawsuit," revealed Samuels. "It's a settled case. It was settled earlier this year [2000]. . . . There was another case earlier in the 90's, 1996 or 1997 . . . and as we began to launch this retesting program we received a third lawsuit." Samuels explained that of the approximately 36,000 tests administered between 1992 and 1998, "we are aware of three [Tay-Sachs babies who were born]."[23]

Two Tay-Sachs Babies

"I was sad to hear what happened, but I wasn't surprised," reflected Shari Ungerleider, referring to Quest Diagnostics' Tay-Sachs retesting program. The Ungerleiders were tested for Tay-Sachs years ago, and because of physician error they were told they were not carriers. Shari subsequently gave birth to Evan Lee, who was a Tay-Sachs baby. Ungerleider explained that she knows of one couple who was informed by Quest Diagnostics that they were not carriers. That couple's son—born five years ago—was also afflicted with Tay-Sachs.[24]

"The doctor tested me [for Tay-Sachs] and misread the test and told me I wasn't a carrier," recalled Shari Ungerleider. "After Evan was diagnosed with the disease, I was retested and my husband was tested and we were both carriers." Evan Ungerleider died of Tay-Sachs disease at the age of four.[25]

When Shari Ungerleider became pregnant for the second time, she had a CVS test (chorionic villus sampling, a prenatal test done early in the pregnancy), which indicated that the baby boy was unaffected. Her third pregnancy, screened at ten weeks by CVS, was found to be positive for Tay-Sachs disease and was terminated. However, the Ungerleiders went forward with a fourth attempt, and their baby girl, whose prenatal test was negative for the disease, was born in late 1999.[26]

The Jarashow family lost their son Noah to Tay-Sachs in November 1999. Jonathan Jarashow's book *The Silent Psalms of Our Son* reveals how this heartbreaking disease can devastate family members, who, after three, four, or five years of watching and waiting for their child to die, search for meaning in that experience. Jarashow and his wife, Judi, cared for Noah in their home until he died just after his fourth birthday. Jonathan Jarashow recounts his dreams of Noah's spiritual life. "Before we found out about his condition, Noah was my earthly child for eight months," confides Jarashow. "During that time I placed so many hopes and dreams in my son. When I look at pictures of him at that age and before, it especially breaks my heart. . . . I didn't realize that he had a very short heavenly string attached to him. I try to avoid these pictures at times . . . for they still break my heart all over again."[27]

Addressing his son, Jarashow reflects, "Of course, on the outside, people only see your physical attributes in this world—that you can't walk and talk. They don't see the side of you that runs around, and sits on God's lap in heaven."[28] Proceeds from Jarashow's book benefit Noah's Spark Foundation for terminally ill children. That charity, as well as the Evan Lee Ungerleider Foundation of NTSADA, serve as additional vehicles for these grieving parents to find meaning in their beloved children's short lives.

The Effects of Canavan Disease

"They are not able to do anything physically for themselves," explained Orren Alperstein Gelblum, describing some of the effects of Canavan Disease on children. Gelblum, who lost a child to Canavan Disease, is president and a founder of the Canavan Foundation in New York City. "Some children seem to develop a little bit in that direction, and then lose whatever they may have developed," Gelblum continued. "[They have] very little motor coordination. . . . Our experience is that they are severely retarded children. Some people claim that their [affected] children do understand some language. Our experience is that it's minimal . . . and speech is nonexistent."[29]

"Our first child was perfectly healthy, and then four years later our second child was born," related Gelblum. "It was clear to us early on that something was wrong, but it took us over a year to find out exactly what it was. And at that time we really were not comfortable trying to have additional children because we didn't know if the prenatal testing would be accurate."[30]

In 1993 scientists discovered that Canavan disease is caused by a defect in the ASPA gene, which produces the enzyme aspartoacylase. In normal brain cells aspartoacylase breaks down a chemical called N-acetylaspartic acid, or NAA; so when aspartoacylase is missing, NAA accumulates and causes brain damage. Excess NAA causes destruction of myelin, an insulating material in parts of the brain and in nerves. When myelin is damaged, nerve cells, or neurons, become impaired in their ability to integrate messages within the brain and in transmission of messages between the brain and the body. For instance, visual signals from the eye are not transmitted properly to the brain, so children with Canavan disease progressively lose their sight. Most Canavan children retain acute hearing, though, since the nerve which transmits signals from the ear to the brain is much shorter, and does not depend as much on myelin.[31]

The test for carrier status relies on detection of a specific mutation, or error, in the DNA. For Ashkenazi Jews, more than 97 percent of Canavan cases are due to one of two possible mutations (termed E285A and Y231X).[32] Therefore, screening for those two mutations can identify most carriers in the Ashkenazi Jewish population. In the general population Canavan can be caused by any one of more than seventy different mutations, thus it is much harder to screen for. It is also much less prevalent in the general population, although the precise percentage of carriers has not been calculated.[33]

The test for Canavan carrier status involves PCR analysis of DNA samples extracted from blood. PCR, or polymerase chain reaction, is a technique that provides for the analysis of genes from a very small sample of DNA. PCR is used routinely now in forensic science, since small samples of blood, semen, and hair, which are typically found at crime scenes, can provide enough cells to reveal the identity of the perpetrator. According to the Mount Sinai Genetic Testing lab report on my own Canavan test, "PCR amplification and restriction analyses revealed that Miryam Wahrman had normal alleles at the two mutation sites. In individuals of Ashkenazi Jewish descent, testing for the above 2 mutations identifies approximately 97% of CD alleles. A negative result reduces the carrier risk from about 1/40 (Ashkenazi Jewish population risk) to about 1 in 1300. . . ."[34]

"Children with Canavan Disease seem to die in their first or second decade of life. Occasionally they may last a couple of more years," explained Gelblum. "They don't all die within the first five years, as do children with Tay Sachs."[35]

"The course of the disease really varies and because it's so variable, some of the children are in and out of the hospital. . . ," she continued. "They tend to have urinary tract infections, pneumonia . . . some children who live a little bit longer develop scoliosis which puts pressure on the heart . . . they have feeding problems, and seizures are fairly common in this disease. . . . Some children are institutionalized, and some are not. Our daughter lived at home. She died when she was seven and a half. Some people's children have lived at home for a while, and as the children got older and more difficult—physically— to handle, to lift, they were placed in institutions close by."[36]

"Care of a Canavan child can be completely time consuming—all consuming," reported Gelblum. "While they can be very happy, cheerful children, they can also be very fussy and difficult and so it can be a very difficult experience."[37]

Familial Dysautonomia

As the Human Genome Project continues to yield data relevant to the understanding of many traits and diseases, individual labs are busy ferreting out important genes. The Human Genome Project, which involves work by scientists in labs in many countries, is working on the mapping, sequencing, and analysis of all the genes carried by human cells. One such gene is IKBKAP; a mutation in this gene is responsible for the disease familial dysautonomia (FD). Two groups reported the discovery of this mutation in the March 2001 issue of *American Journal of Human Genetics*. Research in genetics has progressed to the point where this type of undertaking can require a massive investment of scientific resources. In fact, one group of eighteen scientists, headed by Dr. James Gusella, completed its research in nine separate labs, including Massachusetts General Hospital, Harvard University, Hadassah University Hospital in Jerusalem, the National Human Genome Research Institute, and the Max-Delbruck-Centrum for Molecular Medicine in Berlin.[38] The second, smaller group, headed by Berish Rubin, completed its work at Fordham University and was funded by Dor Yeshorim, the Committee for Prevention of Jewish Diseases (Brooklyn, N.Y.). [39]

According to these reports, the familial dysautonomia gene, located on chromosome #9, has a mutation that results in a missing segment in the mRNA—a molecular copy of the DNA code that is used as a blueprint in building protein. Interestingly enough, the errors do not occur in all cells; "the deleted message is seen in RNA isolated from brain." This is consistent with disorders of the nervous system, which are characteristic of the disease. The gene does appear to produce normal RNA in some cells. This gives researchers hope that they may discover how to correct the defect in nerve cells. As Dr. Susan Slaugenhaupt explained, "The fact that we see normal protein expressed in some patient cell types, despite the presence of the mutation, is unusual and important. . . . We may learn how to increase the amount of normal protein made in nerve cells, which could lead to a treatment for the progressive neuronal loss seen in FD."[40]

In addition, the research showed that "the ethnic bias is due to a founder effect, with > [more than] 99.5 percent of disease alleles sharing a common ancestral haplotype."[41] This means that almost all of the genes that cause this disease in Ashkenazi Jews have exactly the same mutation, probably inherited from a single ancestor—the person in whom this genetic error first appeared. When this person's descendants who are carrying the flawed gene meet and mate, the combination of two copies of this unique mutation is deadly.

The gene responsible for familial dysautonomia is carried by an estimated 1 in 30 Ashkenazi Jews[42] (as compared with 1 in 25 for Tay-Sachs, 1 in 25 for CF, and 1 in 40 for Canavan[43]). This would translate into a frequency of 1 in 900 Ashkenazi couples where both members are carriers, and 1 in 3600 chance for an FD child to be born to an Ashkenazi couple (since a couple where both parents are carriers has a 1 in 4 chance of conceiving an FD baby).

Babies born with FD have nervous systems that do not develop properly. According to information from the Dysautonomia Foundation, "They have trouble swallowing and sucking, and 60 percent require tube feedings. The diagnosis of this baffling disease is often missed and treatment delayed. Patients also have problems controlling their heart rates and blood pressure and they have no taste buds or tears. Despite improvements in medical treatment, only about 50 percent of patients live to be 30 years old."[44]

In cystic fibrosis, a typical screening might involve looking for any of ten different mutations; The Mount Sinai Genetic Testing lab report explains that "testing for these mutations detects approximately 85 percent of CF chromosomes."[45] In Familial Dysautonomia, on the other hand, only one mutation needs to be identified in order to flag more than 99 percent of Ashkenazi carriers. Since there is one version of the mutated gene that occurs in most FD patients, it was possible to develop a highly accurate carrier screening test; such a test has now been added to the panel of tests for the seven other deadly diseases. This development will enable anyone in the Ashkenazi Jewish population to determine carrier status for FD and, just as carrier testing did for Tay-Sachs, it should significantly reduce the number of FD babies who are born.[46]

To Screen or Not to Screen?

Is it permissible for couples to be screened in order to prevent the birth of a baby with a genetic disorder?

One serious problem associated with screening the population for disease genes is, according to Rabbi Moshe Tendler, "that diagnosis without therapeutic potential is hurtful, harmful, and should not be available." He explained that Jews "do not hurt an individual in order to benefit society." Tendler opposed screening more than twenty-five years ago during those first screenings for Tay-Sachs at

Yeshiva University, and he still opposes such programs, since he does not support the option of abortion of an affected fetus. Tendler supports the view of British surgeon S. P. Parbhoo, who wrote in *The Jewish Chronicle* [of London], "We strongly urge all women, whether Ashkenazi or not, to decline any invitation for genetic screening."[47]

On the other hand, Rabbi J. David Bleich, who also opposes abortion of a "physically malformed or mentally deficient" fetus, does accept screening before marriage. "Blood-testing programs as a screening method for the identification of carriers of Tay-Sachs disease before marriage pose no halakhic problem." Bleich cites Rabbi Moshe Feinstein as supporting testing before marriage. Bleich concludes that "marriages which are likely to lead to the birth of defective children should not be encouraged." But if the two are married, they are then obliged to procreate, even if there is a risk of serious genetic defect.[48]

Bleich cites Rabbi Eliezer Waldenberg in an earlier article. "Rabbi Waldenberg rules that abnormality of the fetus is sufficient justification for termination of pregnancy within the first trimester provided there is as yet no fetal movement."[49] In 1977, when Bleich published this article, there was no technology for screening so early in the pregnancy. (Amniocentesis, which was the only recourse at that time, is performed at fifteen or sixteen weeks of gestation). Now chorionic villus sampling (CVS), and a more recently developed "early CVS" can provide information on fetal genetics early enough to permit abortion in the first trimester.

Some issues of concern regarding mass screening of Jewish singles are presented by Dr. Fred Rosner. "Mass screening programs may produce a psychological burden on those young people who screen positive. Should a carrier of the Tay-Sachs gene refuse to marry a mate who has not been tested? Should two carriers break an engagement if they learn that both are carriers as a result of a screening program? Should a young person inquire about the Tay-Sachs status of a member of the opposite sex prior to meeting that individual on a social level? Must a person who knows he or she is a carrier divulge this fact to an intended spouse? Is primary prevention of Tay-Sachs disease by mate selection the proper Jewish approach?"[50] These issues can introduce additional stress into the Jewish dating scene, whether it involves modern young Jews making their own matches in college or the workplace, or more traditional Jews meeting through a matchmaking service.

Bleich explains that "Since a Tay-Sachs carrier is in no way less healthy than a non-carrier such a child suffers no disadvantage save the advisability of exercising prudence in determining that his or her marriage partner is a non-carrier. Thus there is no reason for any stigma to be associated with the carrier state."[51]

Nevertheless, Rosner points out that people identified as carriers may be stigmatized. "Misinformed or uninformed people may shun and ostracize such carriers. . . ." Thus, strict confidentiality should be maintained in any screening program.[52]

Rosner's conclusion raises the possibility of a new option—preimplantation diagnosis:

If the purpose of Tay-Sachs screening is to provide eligible clients with information and genetic counseling about mating and reproductive options, few will oppose screening. If the purpose, however, is to suggest prenatal diagnosis with the specific intent of recommending abortion of affected fetuses, religious and moral objections might be raised. Preimplantation diagnosis of in vitro fertilized eggs with the discarding of affected zygotes, if any, avoids the issue of pregnancy termination since pregnancy in Judaism does not begin until zygote implantation into the wall of the uterus.[53]

In his book *Be Fruitful and Multiply,* Richard Grazi quotes Rabbi Yitzhak Zilberstein of the Israeli Medical-Halakha Group. "One cannot close the door in the face of despondent people who suffer mental anguish in fear of giving birth to sick children, pressure which can drive the mother mad." Zilberstein further suggests that "in the case of a serious genetic disease which affects the couple," it would be permissible to screen an in vitro fertilized embryo, before it is returned to its mother's uterus.[54]

The use of IVF and preimplantation genetic diagnosis (PGD, also see below) to weed out defective embryos is sanctioned by some rabbinic authorities, partially because the early embryo is considered as *maya be'alma* (mere water) (see chapter 4). In addition, possible psychological trauma of the parents is an important consideration in determining whether embryo selection is permissible. "This permissive attitude towards screening and, implicitly, the discarding of the affected untransplanted embryos rests to a great extent on the lack of standing the embryo has in halakha. However, it is the psychological state of the parents which is invoked rather than any negative quality of life that the child might suffer."[55]

May a Genetically Defective Fetus Be Aborted?

When prospective parents find out they are carriers of any of these se-rious diseases, what course of action is available for them? Such cou-ples could choose not to have biological children, or to use an egg or sperm donor (a noncarrier). This would prevent conception of a child with the disease. Alternatively, these couples could opt for in vitro fer-tilization (IVF) and a technique called preimplantation genetic diag-nosis (PGD); these methods make it possible to screen embryos before they are implanted into the womb so that only nonaffected ones can be chosen. Or couples could go ahead and conceive children, becom-ing pregnant the old-fashioned way. They could screen each preg-nancy to determine if the fetus will be healthy or will have the disease; they may then choose to abort stricken fetuses. Another option is for the couple to bear what God blesses them with. Sometime in the fu-ture such couples might opt to clone one or the other parent. The cloned child would be a carrier, but not have the disease.

In determining a course of action, each couple should consider the severity of the disease they are dealing with, in order to understand the ramifications of the disorder. Three of these diseases are invariably fatal at a young age. There is severe disability and suffering of the pa-tient, disruption and devastation of the family members. The other five diseases are also serious and life threatening, but not universally lethal. Couples have to decide what they can endure in terms of their child's disability and the emotional stress involved in having such a child.

Religious issues and attitudes can help guide the couple in terms of how to handle the situation. This is one area of bioethical decision making where there are dramatic differences between how Jewish de-nominations view the issue. The question of abortion has been dis-cussed by authorities from across the spectrum of religious obser-vance within the Jewish community. While a detailed discussion of the abortion issue is beyond the scope of this book, I will highlight some viewpoints from various sources.

A seminal work on abortion and Jewish law was written by Rabbi David Feldman.[56] Although his book appeared in 1968, a full decade before the first test tube baby was born, Feldman addresses "The Status of the Embryo," explaining that the Talmud considers the fetus "*ubar yerekh imo,*" as a limb of its mother, "rather than an independent entity."[57] His analysis of sources leads to the conclusion that "the foetus is not a person. . . . [O]nly when it 'comes into the world' is it a 'person.'"[58] Feldman also derives, from a variety of sources, "the *following fundamental generalization:* if a possibility or probability exists that a child may be born defective and the mother would seek an abortion on grounds of pity for the child whose life will be less than normal, the Rabbi would decline permission. . . . If, however, an abortion for that same potentially deformed child were sought on the grounds that the possibility is causing severe *anguish to the mother,* permission would be granted."[59]

In an epilogue to the third edition of *Birth Control in Jewish Law,* Feldman emphasizes the "pro-natalist" position of Jewish law with regard to procreation.[60] Feldman maintains that "the rights of women, or reproductive freedom, cannot be admissible as an argument in favor of abortion. . . ."[61] However, he also points out that "abortion is not the same as murder . . . in Jewish law, because murder is one of the three cardinal sins the avoidance of which requires martyrdom." And a fetus in a life-threatening pregnancy is considered to be a *rodef,* "an aggressor against its mother," which may, if necessary, be killed.[62] Feldman also addresses some aspects of "New Reproductive Technology," including multifetal pregnancy reduction (MPR), concluding that some fetuses of a multifetal pregnancy may be eliminated to ensure the "survival of the mother or the remaining embryos."[63]

An excellent review of the abortion issue in Jewish thought was written by Dr. Marrick Kukin. Kukin points out that the Reform rabbinical assembly supports "abortion on demand" as stipulated in *Roe v. Wade.* Kukin comments that "it should be pointed out that reform Judaism does not use the same set of standards as do its orthodox and conservative co-religionists." Kukin indicates that Rabbi Eugene Borowitz of Hebrew Union College "favors abortion . . . since reform Jews don't use Halachah as their guide. Halachah is only a source of information which they may choose to obey or ignore, depending on what they think."[64]

Rabbi Morrison David Bial points out that Reform Judaism "would allow a woman to have control of her own body as it does with her destiny. A woman with an unwanted pregnancy, especially when she is underage, or the pregnancy is the result of rape or incest or seduction, would be counseled to have it terminated. . . . Nonetheless, abortion is always a major step and never to be entered into lightly."[65]

Rabbi Bernard Zlotowitz, although writing from a Reform perspective, cites two traditional *poskim,* Rabbi J. David Bleich and Rabbi Jacob Emden. Zlotowitz takes issue with Bleich's rejection of screening and abortion of Tay-Sachs pregnancies (see below). Zlotowitz states, "In regard to a child that will knowingly suffer and die, the Gemara is passionate and considerate: abortion is certainly recommended." Zlotowitz also cites Emden's responsum, "permitting abortion merely on the basis of a pregnant woman's *angst.*"[66]

The Orthodox viewpoint is actually somewhat diverse, ranging from nearly complete rejection of abortion to conditional acceptance. Emden, as explained by Bleich, holds the view that fetal abortion is analogous to *hashkhatat hazera* (destruction of the seed). The act of wasting seed (or Onanism) is prohibited when it is done *levatalah* (for no purpose). However, Emden considers cases where there is a "grave need" as being purposeful, hence under those conditions abortion—at any stage of the pregnancy—may be permitted.[67]

Bleich stipulates that "amniocentesis carried out solely for the purpose of diagnosing severe genetic defects such as Tay-Sachs disease serves no therapeutic purpose. Since the sole available medical remedy following diagnosis of severe genetic defects is abortion of the fetus, which is not sanctioned by Halakhah in such instance, amniocentesis, under these conditions, does not serve as an aid in treatment of the patient and is not halakhically permissible." In fact, Bleich declares that such a procedure may impose a needless medical risk to fetus and mother. He suggests that fetal screening of this sort "also constitutes an act of *chavalah*—an unwarranted assault upon the mother."[68]

Rabbi Moshe Feinstein "strongly condemns abortion for Tay-Sachs disease and even questions the permissibility of the amniocentesis that proves the presence of a Tay-Sachs fetus, since amniocentesis is not without risk, albeit small." However, Rosner reveals that, according to Feinstein, "extreme emotional stress in the mother leading to suicidal intent might constitute one of the situations in which abortion would be sanctioned."[69]

Rabbi Shlomo Zalman Auerbach, an esteemed Israeli rabbi, "prohibits the abortion of even a Tay-Sachs fetus." (Interestingly enough, he permits selective abortion of some fetuses in the case of multiple pregnancy, but only if the lives of the mother and other fetuses are in danger).[70]

According to Dr. Avraham Steinberg, Rabbi Eliezer Yehudah Waldenberg takes a much more expansive view of the issue. "Abortion is forbidden when there is no valid medical reason to allow it. The prohibition against abortion, however, is only rabbinic," explains Steinberg. "Abortion is certainly not considered murder." Steinberg lists situations where Waldenberg might permit therapeutic abortions. They include: "A substantial risk exists that the fetus will be born with a deformity that would cause it to suffer," and, "If it has been conclusively proven (i.e., by amniocentesis) that the fetus has Tay-Sachs disease . . . Down's syndrome or other disorders causing severe mental deficiency."[71]

A Conservative Jewish viewpoint posited by Rabbi Seymour Siegel of the Jewish Theological Seminary is described by Kukin. Although Kukin states that Siegel "takes a middle ground, being more liberal than most orthodox thinkers and more conservative than his reform colleagues," Siegel's view appears to mirror that of a number of Orthodox rabbis. "Siegel is staunchly against abortion on demand . . . [but] allows for abortion in cases of great need."[72]

And surprisingly, Zlotowitz, presenting from a Reform Jewish perspective, tempers his approval of fetal screening and abortion: "This Responsum sanctions genetic testing for the cases cited here only. It is not a blanket approval for all genetic testing. If other cases arise, such as genetic testing to manipulate genes to have children that are geniuses or have desirable physical characteristics, etc., that would have to be considered separately."[73]

Screening Options

The types of fetal testing available to determine genetic status of the fetus include chorionic villus sampling (CVS), which is done at ten to twelve weeks, and amniocentesis, done at fifteen to sixteen weeks of gestation. The so-called "early CVS" addresses the needs of ultra-Orthodox Jews; it is a variation on the conventional CVS technique. "Early CVS has been done among the ultra religious as an avenue for termination," explained Zinberg. The idea behind this new development is to terminate the pregnancy within forty days post-conception, since destroying an embryo less than forty days old is not considered by halakha to be a destruction of human life. The availability of early

CVS, therefore, has provided Orthodox couples with an approach acceptable to many rabbis. If CVS is performed at seven weeks of pregnancy (calculated as time since last menstrual period), which is five weeks or thirty-five days of fetal gestation, then it is possible to get the results, and abort if necessary, before forty days of gestation. However, admitted Zinberg, "[Early CVS] is technically more difficult and there is more risk [to the fetus] involved. The numbers [of these cases] are still small so we can't give the percent risk to the pregnancy."[74]

Another important development, called preimplantation genetic diagnosis (PGD), involves testing the embryo before it is implanted in the uterus. It can only be done in conjunction with in vitro fertilization (IVF). Preimplantation diagnosis involves microsurgery on the embryo a few days after it is conceived in a petri dish in the IVF laboratory. One cell from each embryo is removed and genetically tested for specific genetic disorders. Then, only the embryos that are found to be unaffected with those disorders are implanted into the mother's uterus. This technique obviates the need for abortion, since an embryo with the genetic disorder would never be implanted into the mother's womb. Its major drawback is that it involves IVF, making the technique expensive and technically difficult to accomplish. Thus PGD is still considered highly experimental. Chapter 7 discusses embryo screening in greater detail. The possibility of curing genetic disorders is also discussed.

Decisions regarding genetic diseases are never going to be easy. One parent who experienced the heartbreak of a seriously ill child commented on the choices available. "I do believe that people should have the information and should know realistically what it is they're facing," advised Orren Alperstein Gelblum. "It's a very private decision that people make who discover that they're carriers. I can tell them the very complex, difficult experience it could be to have a Canavan child, but the decision of what to do is a very private one." Orren Gelblum has experienced the trials and tribulations of living with, loving, and losing a Canavan child. "If I were in the position of being pregnant and I knew I was carrying another Canavan child, I would not choose to have another Canavan child," confessed Gelblum. "But that doesn't diminish the experience that I had with my child who passed away."[75]

CHAPTER 7

Designer Genes, Designer Kids

WHEN I FIRST held a "test-tube baby" in my hand eighteen years ago, I was moved by the human potential in that tiny speck in the petri dish. I realized then that in vitro fertilization (IVF) would become an important technique not only to help infertile couples, but also eventually to improve on nature, perhaps by curing genetic disorders.

I was part of a group headed by Dr. Jon Gordon of Mount Sinai Medical Center. Gordon had already established himself as a pioneer in transgenic biology, which in the early 1980s involved injecting DNA into mouse embryos to alter their genes. So in addition to setting up a brand new human IVF lab (the first in New York), we were also performing DNA microinjection and sperm microinjection experiments on mouse eggs. These types of experiments helped to pave the way for human embryology work.

Fertilization occurs when sperm meets egg; sperm penetrates egg; and development is initiated. This stepwise process can be observed in the microscope during in vitro fertilization. Extensive research has been done to reveal the biochemical interaction between sperm and egg. A sperm cell binds to and fuses with the egg membrane; only one sperm cell is permitted to enter. Next, the sperm nucleus, containing its twenty-three chromosomes, undergoes changes to form a male pronucleus—the package of paternal chromosomes found in the fertilized egg. The egg chromosomes complete their meiotic division, which reduces the number of egg chromosomes from forty-six to twenty-three; the female pronucleus, which is the package of maternal chromosomes found in the fertilized egg, is formed.

Once the zygote, or fertilized egg, is generated, the male and female pronuclei, two genetic compartments containing the father's and mother's chromosomes, respectively, are clearly visible. Some hours later the separate compartments disappear, allowing maternal and paternal chromosomes, once again totaling forty-six, to intermingle for the first time. The new genetic combination for this embryo has been established. The zygote will copy all its DNA (DNA replication), then undergo a round of cell division called mitosis. Two cells are now present; those cells, called blastomeres, are genetically identical to each other, and both cells carry the full complement of forty-six chromosomes. Another round of DNA replication, followed by cell division, results in a four-cell embryo. The DNA in each cell is copied and, when the four cells divide again, an eight-cell embryo results. All cells are still genetically equivalent, and all hold the instructions, and even the potential, to become a full human being. The cells of the very early embryo are still totipotent, which means that any of them could theoretically become a whole human being. In fact, if a two-cell embryo falls apart and both cells go on to develop independently, then monozygotic twins are produced. (Monozygotic refers to their origin from one zygote, or fertilized egg).

The very early embryo—a tiny cluster of cells—clearly has enormous potential. And now that it can be manipulated, dissected, and studied at the level of the genes, it is possible to predict what that potential is. Micromanipulation of eggs and embryos has led to the development of a variety of techniques to control reproduction and to screen for genetic disorders while the embryo is still outside the woman's body—before implantation even occurs.

Micromanipulation involves the use of microscopic tools. Some tools are designed to hold human eggs, which measure slightly larger than one-tenth of a millimeter in diameter.[1] Some are designed to hold sperm cells. Microscopic needles serve to penetrate eggs, either for shoving in a sperm or for injecting small (picoliter quantities) amounts of DNA. Other tiny tools can deliver a stream of chemicals to the egg. When experimentation on eggs and embryos began, it was not known whether disrupting the integrity of the egg would result in abnormalities or birth defects.

Helping Sperm Penetrate the Zona

When sperm are unable to penetrate eggs, the eggs can be microsurgically altered (as discussed in chapter 4) by chemical digestion of the zona pellucida (the clear shell surrounding the egg), or by using laser beams to cut through the zona, or by physically tearing a hole in the zona. All of these methods have been applied with varying success.

It is no surprise that Israeli clinics have been at the cutting edge in infertility research. Children and family life play central roles in Israeli society; so infertile couples are willing to go to great lengths to conceive. In addition, the national health care system covers in vitro fertilization for a couple to have two children. (The procedures involved in producing two IVF babies would cost an American couple tens of thousands of dollars.) Those factors have led to the development of more than two dozen state-of-the-art IVF clinics in Israel—an extremely high number for a population of approximately 6 million.

Some of the Israeli clinics are working on developing valuable new techniques to help couples conceive. One such technique involves using laser beams to help eggs hatch out of their protective coat—the zona pellucida. Scientists have observed that eggs from women who had repeated IVF failures and eggs from older women sometimes looked different because the zona pellucida appeared thicker. Experiments involving the thinning out or rupture of the zona became known as "assisted hatching," because once fertilization occurs the embryo must hatch out of its zona in order to implant into the uterine lining.

The earliest methods to hatch eggs involved the use of either acid solution to erode the zona or a sharp needle to tear a hole in the zona. A third method, developed by doctors at Hadassah Medical Center in Jerusalem, used laser beams to burn away a part of the zona.

"There is a segment of recurrent patients who produce eggs with a thick zona, or defective embryos with regard to dissolving the zona," reported Professor Neri Laufer, Director of the In Vitro Fertilization program at Hadassah Medical Center on Mount Scopus in Jerusalem. He explained that the problem with mechanically piercing the zona is that frequently the ruptured zona heals, and the embryo cannot emerge. This tearing method also was reported to be associated with an increased frequency of monozygotic, or identical, twinning. And that might indicate that the embryo itself is being disturbed or altered, since formation of identical twins involves cells of the early embryo pulling apart from each other. Artificial twinning is not desirable, because implications for the future health of the babies are unknown.[2]

Laufer's team has developed a method to open a window in the zona using laser energy. The group built a prototype of a miniature laser, which "acts by photochemistry to dissolve the chemical bonds in the shell," explained Laufer. "You need about ten zaps with the laser to etch a hole big enough for the embryo to hatch." In order to prevent genetic damage the laser is set at a wavelength that will not damage DNA. Laufer described how the machine is similar to the one used in keratoplasty—corrective surgery of the cornea of the eye—so it is approved for use in the United States for that application.[3]

Laser-assisted hatching appears to increase success in IVF of younger woman who have suffered recurrent IVF failures. However, in a group of patients where the women were older than thirty-eight, the technique did not seem to increase the success rate. "They produce eggs that *a priori* are defective," concluded Laufer.[4]

Rabbinic concerns about this type of technique might include the issue of *hashkhatat hazera*—improper emission or destruction of the seed. However, since the intention of this technique is clearly to improve the likelihood of reproductive success—and not to destroy the eggs—that should not be a concern.

Cryopreservation: Freezing Eggs

Once the egg is fertilized, the zygote or embryo can be manipulated. Four- to eight-cell embryos can be cryopreserved, that is, frozen for future use. The advancements in cryopreservation have enabled couples to use the embryos generated in one cycle to attempt future pregnancies. Couples can also consider donating frozen embryos. Or researchers can save and utilize frozen embryos in a wide array of research projects to further our understanding of the early stages of developmental biology.

The ethical issues that arise due to cryopreservation are complex, since cryopreservation provides options not previously available. In IVF, the woman is given hormones that stimulate the ovaries to produce more than one egg per cycle. Patients typically respond to hormonal stimulation by producing several eggs. In some cases, as many as ten or twenty eggs can be produced. Thus the question arises—what do clinics do with the surplus eggs? Generally all eggs are used in the next step, fertilization with the partner's sperm. Depending on the success of fertilization, the couple may end up with five, ten, or even as many as twenty embryos. Usually only three embryos are returned to the mother's womb in any given cycle. The remainder can be frozen away for future use. If the cycle is successful and a baby (or babies) is born, the couple can elect to destroy the remaining embryos, or donate surplus embryos to another infertile couple, or use them for future cycles to add to their family.

A number of issues arise, some of which have already been discussed above in chapter 2. For instance, if embryos can be frozen, they can be donated to just about anyone. In the case of donated embryos, how do we identify who are the parents? Are they the genetic parents (the embryo donors) or the recipients?

Cryopreservation can lead to donations within the family (sisters donating to sisters, daughters donating to mothers, etc.). Are there problems involved in terms of prohibited family relationships? Can embryo donations result in *mamzerim* (illegitimate offspring; also see chapter 3) being born? One problem that may arise is that the children of donated embryos could unwittingly meet and marry their genetic siblings.

A recent news report discussed a couple who had become parents through IVF. They felt that their family was complete, and they were left with twenty-three surplus frozen embryos. They turned to a program called the Snowflake Embryo Adoption Program, which matches up donor and recipient couples. The unique aspect of this program is that the donor couple can meet and interview the recipients. They can even arrange to have an ongoing relationship with their genetic child, if both couples are agreeable to that arrangement. Their children can get to know and have a relationship with their genetic siblings as well. These types of "open adoptions" may actually be preferable from a halakhic perspective, since it would eliminate the possibility of siblings marrying each other unknowingly.

Frozen embryos have been donated to women who are too old to generate their own eggs. There have been cases of women in their sixties giving birth to babies as a result of adopting eggs and embryos. Are there ethical problems with this scenario? It would seem that an older woman would be exposed to higher levels of risk in pregnancy than a younger woman. Is it permissible to expose oneself to unnatural levels of risk in order to give birth? The issue of *pikuakh nefesh*—preservation of life—may be critical in this type of case. If the woman is healthy and doctors judge the risk small, then it may be permissible. If doctors are concerned that she may experience serious complications, then there could be halakhic questions about using such technology to produce a pregnancy in older, postmenopausal women. There is biblical precedent for an older woman, well beyond her fertile years, conceiving and delivering a child. Sarah and Abraham conceived very late in life. Upon hearing God's prophecy regarding the birth of Isaac, "Abraham threw himself upon his face and laughed, and he thought: 'Shall a child be born to a hundred-year-old man? And shall Sarah—a ninety year old woman—give birth?'" (Gen. 17:17).

The ability to freeze embryos has also led to bitter custody battles involving couples who went through IVF, had surplus embryos frozen, then divorced. These types of cases touch upon a person's fundamental right to be a parent or not to be a parent. In one such case the father wanted to use the embryos in a future relationship or to donate them to an infertile couple. The mother wanted the embryos destroyed or used for research. A Superior Court decision ruled in favor of the mother. Since then, the decision has been appealed. An earlier case in Tennessee led to the ruling that a man's right not to have children outweighed his partner's right to donate their embryos to an infertile couple.

In yet another case, involving a couple in Israel, the wife (who had a hysterectomy) wanted to have the surplus embryos gestated by a surrogate mother, while the husband wanted them destroyed. The district court ruled in her favor; the high court ruled against her. The religious court ruled to stop the IVF procedure, since a surrogate mother was involved and the religious courts did not, at that time, sanction surrogacy. (Since then religious courts have, in some instances, permitted surrogacy arrangements.) In order to avoid future litigation between "parents" over custody of frozen embryos, most IVF programs that offer cryopreservation require that the couple complete and sign a form stipulating how they wish any surplus embryos to be used.

Halakhic views regarding the different options involved are fairly complex (also see chapters 2 and 4). Rabbi Yitzchok Breitowitz summarizes some of the issues in his 1996 article:

At one extreme, most contemporary posekim . . . have allowed the destruction or at least the passive discarding of "unwanted" preembryos, ruling that the strictures against abortion apply only to embryos or fetuses within a woman's womb and not to preembryos existing outside of it. Experimentation on preembryos not destined for implantation would appear to be permitted as well. By contrast, embryo donation to infertile couples—whether Jewish or non-Jewish—or to a single woman, raises numerous halakhic and ethic complexities and has not to date received widespread halakhic sanction. (Indeed, in a worst-case scenario—embryo donation to a married Jewish couple—such donation may even constitute the commission of halakhic adultery and result in the birth of *mamzerim*).[5]

Preimplantation Gene Diagnosis

Another issue that raises halakhic questions involves the manipulation and screening of fertilized eggs and early stage embryos. An early embryo can be pulled apart and individual cells can be studied for diagnostic or research purposes. In preimplantation genetic diagnosis (PGD), an eight-cell embryo can be screened for genetic disorders. The microscopic clump of cells is held in place with a micropipette. A small tube is used to suction off a single cell, leaving seven intact cells. That single cell contains the same genes as the other cells, and by analyzing the DNA or the chromosomes in the single cell it is possible to get a picture of the genetic complement of the embryo. This process is analogous to taking a blood sample from an adult and performing a DNA test on the white blood cells. However, with PGD the test must be done on a much smaller scale and can rely on polymerase chain reaction (PCR) technology, involving the amplification of very small quantities of DNA before analysis.

The seven remaining cells are not missing any crucial parts (you do not produce a baby missing one limb because one cell was destroyed at the eight-cell stage). The seven cells are totipotent, thus they simply reorganize and compensate for the loss of one cell. PGD can identify which embryos have genetic disorders such as Tay-Sachs, cystic fibrosis, and Canavan disease. Or the chromosomes can be studied by a procedure called fluorescence *in situ* hybridization, or FISH. That procedure takes advantage of the fact that DNA is double stranded; the DNA code (in the form of chemical bases) on one strand binds to that found on the other strand. Thus, if you take a small piece of DNA whose genetic code matches part of a chromosome, it will bind to that chromosome. Such DNA fragments can be tagged with fluorescent dyes, and when the tagged DNA pieces bind to chromosomes, the chromosomes appear as bright spots in the nucleus of the cell.[6]

Using this method it is possible to determine if an embryonic cell is carrying the normal number of chromosomes (i.e., 46) or an abnormal number. Embryos with extra or missing chromosomes can result in offspring with serious genetic defects. (An extra chromosome 21, for instance, causes Down syndrome.) FISH is usually used to check cells for the most common chromosome abnormalities in human embryos—the presence of an extra chromosome 13, 18, 21, or abnormal complements of sex chromosomes.[7] This approach enables a couple who produced several embryos in IVF to select those with normal chromosome complements for implantation in the uterus. It permits genetic selection without the trauma or halakhic questions involved in fetal abortion.

One case utilizing PGD involved a couple who had already experienced two pregnancies with fetuses testing positive for Tay-Sachs. The couple had chosen to abort the fetuses rather than lose the children to that horrific disease. Instead of risking another Tay-Sachs pregnancy, they turned to PGD. The couple produced four embryos in an IVF procedure at the Jones Institute for Reproductive Medicine in Norfolk, Virginia. A single cell was removed from each embryo, and the DNA was analyzed to determine which embryos were free of the disease. Two turned out to be carriers, and the other two were completely free of the Tay-Sachs gene. All four were implanted into the mother's uterus; one embryo developed, and the little girl who was born was not even a carrier of Tay-Sachs. PGD allowed the couple to have a healthy child without the fear of another aborted pregnancy. Although PGD is an expensive technique, because it involves all the steps of IVF, plus the additional steps of microdissection and DNA analysis, the more than $10,000 cost may be well worth it for couples who would not consider aborting a Tay-Sachs pregnancy.

Another case, reported in *The Jerusalem Post* in 1998, was even more exceptional. The "miracle baby" who was born in Hadassah University Hospital was fathered by a man with Klinefelter syndrome. Males with Klinefelter have an extra X-chromosome—they are XXY—and they do not produce mature sperm in their semen. The

Hadassah doctors were able to recover two live sperm from the man's testicle; they used them to fertilize two of his wife's eggs in vitro, and both embryos were tested by preimplantation diagnosis to determine if they were affected by the disease. Chromosome tests showed that one embryo was normal (XY) and the other had the XXY chromosome complement of Klinefelter's. Only the normal embryo was implanted. The baby boy who was born was the first Israeli baby to be born to a sterile man with Klinefelter syndrome.

Is Embryo Splitting Permitted?

Once you are able to manipulate the embryo and dissect it, pulling cell from cell, this can lead to "embryo splitting," a technique used routinely in cattle to produce multiple copies (twins, triplets, or quadruplets) of cattle with desirable traits. If done with humans, the result is monozygotic or identical twins, triplets, or quads, as well. Embryo splitting is one form of cloning that occurs in nature. Although it is uncommon, most of us know people who are identical twins (you may even be a twin yourself). However, is it permissible to produce twins deliberately? Is it desirable to produce identical twin, triplet, or quadruplet babies artificially from one embryo?

In the discussion above on assisted hatching, it was noted that some of the manipulations of the egg and its membranes may lead to an increased incidence of monozygotic twinning. An article from 1996 indicated that zona manipulation of eggs, including ICSI (sperm injection), zona drilling, and mechanically assisted hatching, led to higher than expected numbers of monoamniotic twin gestations (where the twins develop in one amniotic sac). The authors' conclusion was that "Monoamniotic multiple gestations may be increased in zona-manipulated cycles. The potential obstetric risks and complications of zona manipulation should be discussed with patients."[8] Although these observations of increased monozygotic twinning have not been universally observed, a handful of studies suggests that tampering with the zona will result in some unintended consequences, including artificial embryo splitting.

It is conceivable that some couples might desire deliberate embryo splitting. For instance, if their IVF cycle yields a low number of eggs, it might increase the chance of success if their embryos are split and then replaced in the uterus. Infertile couples generally accept (even welcome) a risk of twins, and even triplets, if the alternative is failure to conceive. If we deliberately induce the formation of identical twins, we are cloning humans. This is different from the cloning of people who were already born (as discussed in chapter 5). But it is cloning, nonetheless, since cloning is defined as production of genetically identical individuals. The deliberate induction of twinning is closer to what can occur in nature when identical twins or triplets develop.

From rabbinic writings on cloning and on reproductive issues, it would appear that embryo splitting would be halakhically permissible. After all, experimentation on the preembryo is permitted by most rabbis. The injunction for procreation is being fulfilled (even more so, if more babies are born as a result). Thus, it seems there would be no halakhic barrier against embryo splitting.

And yet, there may be a "yuck factor" (a level of disgust generated by an unnatural outcome) involved, because the question can arise: How many times may you split the embryo? If twins are acceptable, are triplets? Quadruplets? Quintuplets? Where do you stop? The slippery slope in this case appears to lead to Huxley's *Brave New World* of identical cloned babies.

The halakhic roadblock in this case, preventing a slide down a slippery slope, may be the issue of *pikuakh nefesh*. Embryo splitting—in fact fertility treatments of any kind—may be questionable if the mother's life is placed in jeopardy. And multiple pregnancies, especially those exceeding twins, do increase the risk for the mother. In addition, rabbinic views do not give blanket approval to embryo donation. Thus if embryo splitting is done to produce 4, 6, or 8 identical babies, it would not be permissible to put them all in the same mother (because of the high-risk pregnancy), and it would be undesirable to place them in a surrogate (since the family relationships can be confounded).

Stem Cells

The early embryos can be allowed to further develop to the blastocyst stage—a hollow ball of cells forming about five to six days after fertilization, and stem cells from the inner cell mass (a small clump of cells hanging off the inner wall of the embryo) can be harvested and used to grow a smorgasbord of tissues. Tissues derived from stem cells may be very useful to replace diseased tissue in children or adults. Chapter 4 explores many of the ramifications of this type of embryonic manipulation in detail.

Designing DNA: Gene Therapy

Gene therapy refers to the correction of genetic defects by repairing or replacing flawed genes. In somatic cell gene therapy, cells or tissues of a patient may be removed, new DNA inserted, and the tissues replaced in the body. Alternatively, DNA may be introduced into the patient's body by means of a specially engineered viral vector—that is, a virus that is incapable of causing disease but still retains its ability to "infect" a cell and to bring functional genes into a diseased cell.

Either approach, when used to alleviate or cure a genetic disorder in a patient, would appear to be sanctioned by Jewish law, since it accomplishes the same goals as traditional therapy. "Heal, thou shalt heal" and "*pikuakh nefesh*" could provide license for therapy that saves lives or improves the quality of life. Regarding this point Dr. Fred Rosner wrote, "Gene therapy, such as the replacement of the missing enzyme in Tay-Sachs disease or the missing hormone in diabetes, or the repair of the defective gene in hemophilia or Huntington's disease, if and when these become scientifically feasible is also probably sanctioned in Jewish law because it is meant to restore health and preserve and prolong life."[9]

Rabbi Ira Youdovin of Reform Judaism's Stephen Wise Free Synagogue in New York City expresses surprise at the permissive attitude of Orthodox *poskim* toward many aspects of gene therapy. "The use of genetic testing and engineering as a tool for combating genetically-based diseases is strongly endorsed by a broad spectrum of Jewish opinion, including opinion from Orthodox quarters, where one might expect to find Jewish Luddites screaming *gevalt* over tampering with the work of Creation," reflects Youdovin. "And while there are some Orthodox scholars like Rabbi Moshe Hershler, who argues that 'one who changes the [Divine arrangement] is lacking faith [in the Creator],' Orthodoxy nevertheless tends to accept gene transplants as the analogue of organ transplants."[10]

Some may consider gene transplants as analogous to organ transplants; however, when considering germ-line gene therapy, which involves the altering of genes in eggs, sperm, or embryos, there are unique considerations. The major concern is that any changes in the genes that can be passed on to the next generation will be introducing long-lasting changes into the human population, changes that could be passed from generation to generation throughout the conceivable future. Any mistakes made in this type of gene therapy may have long-lasting effects on the future human gene pool.

Dr. Stephen Modell's article published in the Reform movement's Union of American Hebrew Congregations (UAHC) Program Guide on Cloning expresses the need for caution: "Whatever changes are introduced into the germ line will perpetuate themselves in subsequent generations. For GLGT [germ-line gene therapy] to be acceptable, it must meet higher levels of stringency than the somatic cell gene therapy now being applied to patients, since an error introduced by the procedure will (in almost a biblical sense) visit itself unto the fourth generation (Exod. 34:7) and beyond."[11]

Gene therapy promises to be one of the most exciting applications in embryo research. Gene therapy of embryos involves insertion of DNA from one source into the nucleus of the fertilized egg. In order for the technique to work, the DNA must become part of the embryonic chromosomes and must be capable of being expressed. The three methods most frequently employed to introduce DNA into embryos or other cells are: (1) microinjection of DNA (2) use of viruses to carry the genes, and (3) electroporation—using an electrical impulse to change the permeability of the cell so that DNA may pass through the membrane. They all have advantages and disadvantages; none of these techniques is foolproof. In fact, early experiments with mice produced animals that carried the new genes; however, many of the genetically engineered mice also exhibited abnormalities or birth defects. It is very difficult to ensure that DNA that is brought into the cell actually becomes part of the cell's chromosomes, that it remains active and does not interfere with other genes required for the survival, health, and normal development of the embryo.

"We assume that it has been perfected to the point where it is (almost) always successful," wrote Rabbi Azriel Rosenfeld in one of the first articles on gene therapy and Judaism, referring to genetic engineering of human eggs. "Otherwise, it might be regarded as 'destruction of the seed,' which, according to many authorities applies to a woman's seed as well as to a man's."[12]

On the other hand, Rosner emphasizes that "the sperm or ovum or even the fertilized zygote is not a person. Thus, gene manipulation is not considered as tampering with an existing or even potential human being, since that status in Jewish law is bestowed only upon a fetus implanted in the mother's womb."[13]

Rosenfeld does feel that the obligation to heal also covers the use of gene therapy. "Any surgery that is permitted on a person must certainly be permitted on an ovum or sperm before conception. If a surgical cure for hemophilia were possible, it would surely be permissible; thus it would certainly be permissible to cure hemophilia by gene surgery."[14]

And what about frivolous applications of gene therapy, such as the use of gene therapy to make someone taller, or more attractive? Rosenfeld does not rule out those applications either, reasoning, that "cosmetic plastic surgery is permitted by many authorities in order to relieve psychological distress; they should thus also permit achieving cosmetic effects through gene surgery—assuming, of course that the surgical procedures are safe and reliable."[15]

Rosenfeld even suggested that there are talmudic allusions to changing the quality of one's offspring before they are born. He quotes the story of Rabbi Yohanan who "used to go and sit at the gates of the place of immersion saying: 'When the daughters of Israel come out from their required immersion, they look at me and may have sons who are as handsome as I, and as accomplished in Torah as I.'"[16] That talmudic anecdote suggests that it is possible to change traits of a child before he or she is even conceived. Perhaps it is a sanction for proactive gene therapy to improve the traits of an embryo. (Or perhaps it was just wishful thinking on the part of Rabbi Yohanan, who was apparently quite a handsome as well as a learned man.)

Rosner disapproves of frivolous cosmetic applications of gene therapy. "Should these techniques be allowed to alter human traits such as eye color, height, personality, intelligence, and facial features? If tall basketball players are more successful than short ones, should we produce only tall basketball players? Obviously not. Should we create piano players with three hands? Obviously not."[17]

Youdovin is likewise concerned with where to draw the line in gene therapy. "As a rabbi in New York City, I've seen too many parents go 'bonkers' when their four-year-old doesn't get into the absolutely best kindergarten—which diminishes his/her chances of making it into the

absolutely best college—which ruins the child's life even before it begins losing baby teeth. . . . I'm simply not prepared to guarantee that
potential parents could resist the temptation of visiting their friendly
neighborhood geneticist." Youdovin portrays a scenario where parents would choose the best four or five eggs from several dozen collected over months, to ensure the production of "carefully crafted designer children."[18]

Only a trace amount of DNA is necessary to alter the traits of an
embryo. Picoliter volumes—measured in trillionths of a liter—are injected. "Rabbi Rosenfeld contends that gene surgery might be permissible in Jewish law because genes are submicroscopic particles and no
process invisible to the naked eye could be halakhically forbidden,"
notes Rosner. "Laws of forbidden foods do not apply to microorganisms. The priest only declares ritually unclean that which his eyes can
see."[19] This principle is an important one to keep in mind, not only for
the question of human gene therapy, but also regarding animal and
plant genetic engineering (see chapters 11 and 12).

Rosner also poses a question concerning the origin of DNA. If
human DNA from a relative is used, could this constitute an incestuous act? "Are gene transplants considered to be a type of perverted
sex act between the gene donor and the recipient? Would such transplants be forbidden, in particular, if donor and recipient were close
relatives? Would a child conceived from such a manipulated ovum or
sperm be regarded as related to the gene donor?" Rosner cites
Rosenfeld's argument that, if an ovarian transplant is performed, the
ovary becomes part of the recipient, just as it does for other organs,
such as kidney, cornea, and heart. Offspring of that woman would be
her children and not those of the ovary donor. "Rabbi Rosenfeld
draws parallels from the rabbinic responsa dealing with ovarian transplants and concludes that no sex act is involved in a gene transplant,
the recipient is not forbidden to marry the donor's relatives, and the
child conceived and born following a gene transplant is not related to
the gene donor."[20]

Rabbi Mordechai Halperin discusses the question of whether Jewish law permits research on the human genome; his reasoning with regard to that question should also apply to gene therapy, and may
apply to many other topics addressed here. He cites King Solomon's
proverb, "Happy is the man that feareth always" (Prov. 28:14), which
is not to say that you need to fear all new things, rather, "King Solo-

mon meant to praise the person who is always cautious. Fear is praise-worthy when it motivates caution, when it brings one to check oneself and to make sure that one is not going to fall into a trap, step on a hidden mine, or slide down a slippery slope."[21] So, with appropriate caution, novel technologies are permissible.

Rabbi Shabtai Rappaport supports Halperin's cautious approach to technology. Since the Torah states, "When you build a new house, you shall make a battlement for your roof, so that you should not bring blood upon your house, if any man fall from there" (Deut. 22: 8), Rappaport concluded, "The sages (*Kettubot* 41b) interpreted this verse as a general command to maintain a high level of safety. The act of building the new house itself is not forbidden; we are only commanded that reasonable safety measures be taken."[22]

"Any new technology should be approached with care and practiced under protected conditions," warns Rappaport. "However, we should not fear unforeseeable risks. G-d watches over the world and over man. It is man's duty to obey His orders regarding safety."[23]

Rappaport also emphasized that, although we are not permitted to change the "ordinances of heaven," technology is one of the tools that God gave man to deal with the world. Rappaport maintains, "the halachic viewpoint is that the more advanced the technology, the more reason there is to permit it. Any consideration of an organism being an unchangeable entity has no basis in halachah."[24]

One day in the not-too-distant future, early human embryos could be genetically engineered to correct specific genetic defects. This has been done successfully in some animal experiments. However, the rate of abnormal genetic results in animals has been significant; many therefore believe that applying these techniques to humans at this time would be premature.

Gene therapy of tissues and cells is very difficult and expensive, but it holds out the best hope for curing patients with genetic diseases. Unfortunately, gene therapy is a science that is still in its infancy, and scientists working in this field report only limited success with the approach. There have been hundreds of gene therapy trials at medical research centers worldwide. However, none has yielded an effective cure for genetic disease.

Reports of experimental gene therapy trials for Canavan disease, being performed at Yale University, present hope that one day a cure may be found for that genetic disease. An article in the *New York*

Times Magazine described the saga of two-year-old Jacob Sontag, who was one of the first Canavan children to receive gene therapy—via injections of DNA containing the healthy ASPA gene directly into his brain.[25] The ASPA gene makes an enzyme that is needed to break down an acid that harms myelin, a protective coating on nerve cells. A total of sixteen children ranging in age from nine months to seven years received gene therapy, and, according to the Canavan Research Foundation website, "most have shown clinical improvement."[26]

However, the death of Jesse Gelsinger in September 1999 slowed the progress being made in gene therapy research. Gelsinger died four days after a gene therapy injection in a clinical trial for ornithine transcarbamylase (OTC) deficiency disease. As a result of his death, many gene therapy trials were put on hold.[27]

Since that tragic incident, two reports of successful human gene therapy have appeared. They include a research project in France involving several infants with severe combined immune deficiency disease who improved after gene therapy[28] and a report of two patients with hemophilia B who were injected with the Factor IX gene and began to make their own clotting factor.[29]

Meanwhile, parents of children with Canavan disease eagerly await implementation and results of a new research protocol that holds promise for their own youngsters. In early 2001, the Canavan Research Foundation website happily reported, "Other good news. Rats injected with the vector that the kids will receive are showing motor and cognitive improvements."[30] Reports on human gene therapy trials for Canavan are likewise promising. In the summer of 2001, three American Jewish children with Canavan disease were selected for gene therapy. Because most substances do not readily pass from the bloodstream into the brain, the technique used involves drilling holes into the skull and introducing synthetic viruses carrying the normal human genes directly into the brain. "Doctors report that [the children] all improved on various neurological and psychometric measures."[31]

These initial successes are certain to encourage more aggressive clinical work in gene therapy. As gene therapy becomes more widespread, Youdovin's conclusions should be kept in mind: "It is, however, a matter of repeatedly reminding ourselves . . . that our tradition affirms that humankind was created in the image of God . . . and while that in no way precludes tampering with a human being's DNA—if tampering enhances human life—it surely does preclude tampering with a human being's basic humanity. If we can get this message across . . . it may be enough to head off a catastrophic slide down that ever-threatening slippery slope."[32]

CHAPTER 8

Chosen Children:
Sex Selection

S HOULD WE SELECT the sex of our offspring? The most compelling reason to consider such an action is to prevent the birth of
children with sex-linked disorders, enabling families carrying such
diseases to have healthy children. Some serious genetic disorders, for
instance, almost always strike males. By allowing parents to have
only daughters, couples could avoid having sons with hemophilia, or
Duchenne's muscular dystrophy. Some argue that there is nothing
wrong with using sex selection for "family balancing," that is, the
practice of increasing one's family by adding children of the less represented sex. By that reasoning, a couple with sons should be allowed to choose a daughter, and vice versa. The desire to choose the
sex of one's offspring is not new to Judaism—or to other religions
and cultures. And the question of how to apply sex selection procedures is not a simple one.

The Talmud recommends that if one wants to conceive sons, one
should position the marital bed so that it faces north and south.[1] This
is based on a passage from Psalms (17:14), "whose belly You fill with
tziphuncha; they will have sons in plenty." The interpretation of that
verse depends on the Hebrew word *tziphuncha,* which can carry two
alternate meanings: "hidden treasure" or "north." Another talmudic
prescription to produce boys advises "anyone who desires to beget
male children should contain himself . . . during intercourse, so that
his wife ejaculates [climaxes] first."[2]

None of these ancient techniques has ever been scientifically proven to change the actual birth rate of sons. However, in 1998 medical researchers reported a method that could enable couples to have more control over gender selection in their children. The procedure, called MicroSort—which produces enriched samples of X- or Y-chromosome-bearing sperm—reportedly can produce up to 90 percent females for couples desiring girls and up to 71 percent males are born to those desiring boys.[3]

An even more dramatic development is the application of IVF and preimplantation genetic diagnosis to weed out embryos of the undesired gender. Couples can choose to implant either male or female embryos with virtually 100 percent certainty. Previously this was only done for families who wanted to avoid having offspring with lethal genetic diseases. Now it is even being used when gender is the sole criterion.

The drive to choose the gender of one's offspring goes back thousands of years. In many cultures and religions, including Judaism, males were more highly valued than females. For instance, the Talmud states, "Happy is he whose children are males and woe unto him whose children are females."[4] So the prescriptions for sex selection are usually focused on how to produce male offspring. The Talmud advises, "What should a person do so that his children will be males? Rabbi Eliezer says, 'He should distribute money among the poor.' Rabbi Joshua says, 'He should gladden his wife prior to intercourse.'"[5]

If techniques were readily available to enable parents to choose the sex of their offspring, would it affect the sex ratio? The preference for boys that is apparent in talmudic writings also appears in other cultures, even today. In China, India, and South Korea amniocentesis and ultrasound are used to determine the sex of the fetus. Female fetuses are preferentially aborted, skewing the sex ratios at birth.

Universally, the natural conception rate of male embryos slightly exceeds that of females, so, worldwide, the birth ratio of males to females is approximately 106 boys to 100 girls (without intervention). (It should be noted, however, that since the death rate of males exceeds that of females, girls quickly catch up. In the United States the male to female ratio decreases to 97 males to 100 females (97/100) by ages 25–44, 80/100 by ages 65–69, and to less than 40 men to 100 women by age 85.)[6] In countries, such as China, where selective abortion of females occurs and female infanticide is also practiced, the birth ratio is reported as ranging from 119 boys to 100 girls (in Beijing) to 159/100 (in rural areas of China). In India, ratios reported range from 108/100 to as high as 166/100 in the northern states of Bihar and Rajasthan.[7] In South Korea, rates of 115 boys to 100 girls have been reported.[8]

A 1985 survey in Bombay, India, showed that 90 percent of all amniocentesis procedures performed were done for sex determination, and nearly 96 percent of female fetuses were subsequently aborted.[9] Strict laws were passed in India in 1994 to ban sex determination; those laws provided for steep fines and imprisonment of doctors who revealed the baby's gender. However, "not a single doctor has been convicted," claims Ganapati Mudur. In fact, Mudur reports, "doctors and parents work in collusion, and sex determination involves only oral verdicts by doctors," who therefore cannot be prosecuted since there is no documentation.[10] On one website, <www.howtohaveaboy.com>, a doctor based in Bombay flagrantly advertises sex selection services.[11]

In a 1995 British survey of couples inquiring about sex selection, 83 percent of all couples in the study wanted boys. Ninety-nine percent of Indian couples, and 89 percent of Chinese couples wanted a boy, while 55 percent of European couples preferred a girl.[12] Thus, not all cultures are biased against girls. Sex selection in Europe and the United States would probably have a minimal effect on sex ratios. The Genetics and IVF Institute in Virginia, for instance, reports that requests for boys and girls are about equal.

One recent survey in Japan even shows a bias in favor of females. Apparently, up to 75 percent of young Japanese couples prefer baby girls. Daughters, it is reasoned, are easier to raise, and they are "more likely to look after their elderly parents." However, this pro-female attitude is much stronger among the young Japanese women, and the young men continue to prefer sons. The impact of sex selection techniques on birthrate has thus far been negligible, perhaps because men are reluctant to participate in preferential pursuit of female offspring. The birth ratio has remained unchanged in Japan for the past 100 years, with about 105 boys born for every 100 girls.[13]

Ancient Views on Sex Determination

Many talmudic sources[14] believed that gender was determined at the time of cohabitation. According to the Talmud, the scriptural passage

"If a woman emits her seed and bears a male child . . . " (Lev. 12:2) refers to sex determination. Thus, "Rabbi Isaac the son of Rabbi Ammi said: If the man first emits seed, the child will be a girl; if the woman first emits seed, the child will be a boy."[15] This advice of the Talmud was taken to mean that the woman should climax before the man in order to produce sons. If the man climaxes first, they will produce daughters. (One bonus here for the women is an improved sexual experience, as their men conscientiously attempt to conceive boys.)

One commentary on the Bible wrote the following, regarding Leviticus 12:2: "Some say they found in the Sefer Ha-Teva that in a woman there are seven openings, three on the right, three on the left and one in the middle. If the seed enters on the right side she will bear a male; if [it enters] those on the left she will bear a female; and if [it enters the one] in the middle she will bear a *tumtum* or *androgenes* [hermaphrodite]." Thus, this commentary reasoned, if a woman lies on her right, the seed will enter the right and she will have a son. If she lies on her left side, she will conceive a daughter. This is consistent with a passage from Song of Songs (2:6), "let his left hand be under my head and his right hand embrace me," for that is the recommended position for conception of males.[16]

In addition, in Chronicles it is written: "And the sons of Ulam were mighty warriors, archers; and had many sons, and grandsons."[17] The rabbis asked, "Now is it within the power of man to increase the number of sons and grandsons?" They went on to explain that the sons of Ulam had held themselves back during intercourse to allow their wives "to emit semen first so that their children shall be males."[18] An additional directive to ensure male offspring includes the following: "Raba stated: One who desires all his children males should cohabit twice in succession."[19] Thus, if the man climaxes before his wife, he should climax again after her, to ensure a son. This might be a demanding but worthwhile task for a righteous husband who desires many sons.

Given these views that sex determination occurs during intercourse, the Talmud notes that a man should not pray that his wife will bear a son if she is already pregnant.[20] Since the sex of the child has already been determined by that point, it would be a prayer in vain. However, Julius Preuss discloses that some rabbis felt that such a prayer could still change the sex of the child, as long as it is within forty days of conception. "This would be true in the case where a man and woman emitted their seed simultaneously, so that the fetus is originally sexually neutral," writes Preuss.[21]

Preuss also cites the views of Hippocrates and Galen[22] regarding sex determination, namely that "the right ovary gives rise to boys and the left ovary to girls."[23] Maimonides is said to have considered this view of sex determination but was skeptical that it was possible to determine such a mechanism. Maimonides reacted to this claim, stating: "A man should be either prophet or genius to know this."[24]

Galen also recommended that "one compress and hold fast to the right testicle if one wishes to beget girls," since he believed that "boys arise from the right testicle and girls from the left."[25] Maimonides apparently supported that view. In "Excerpts from Maimonides' Medical Aphorisms" it is written, "One should examine a male at the time he reaches puberty. If his right testicle is larger, he will give rise to male offspring; if it is the left, he will give rise to females. The same situation applies for the breasts of a girl at the time of puberty."[26]

Now we know, however, that women who have had an ovary removed can conceive either a boy or a girl; likewise that either testicle is capable of fathering sons and daughters. In fact, the mechanism of sex determination does not depend on the female "seed" (the egg) at all, it is entirely dependent on the male seed, the sperm.

High-Tech Methods of Sex Determination

Sex selection can be performed in the absence of technology. In China, for instance, female infanticide has reportedly been used.[27] No technology at all is necessary for that approach wherein newborn girls are murdered or neglected and simply left to die. These cases are reported as stillbirths.

Amniocentesis and sonographic fetal monitoring can also be used for sex selection. Amniocentesis involves obtaining a sample of the amniotic fluid surrounding the fetus. That sample contains fetal cells; the chromosomes of those cells can be studied and the sex of the offspring determined. Ultrasound, which takes a sonographic "picture" of the fetus can determine the sex of a fetus with upward of 90 percent accuracy. Armed with information obtained through amniocentesis or ultrasound, selective abortion can occur. These techniques are currently the methods of choice in India, China, and South Korea.

More sophisticated methods depend on being able to preselect for sex of the offspring before conception occurs. The sex of a baby is determined by the joining together of sperm and egg at conception. All eggs carry twenty-three chromosomes, including an X-chromosome. All sperm have twenty-three chromosomes; among the twenty-three chromosomes, half of the sperm carry an X-chromosome and half of them carry a Y-chromosome. When an X-chromosome-bearing sperm fertilizes an egg, it produces an embryo that is XX, or female. When a Y-chromosome-bearing sperm fertilizes an egg, it produces an embryo that is XY, or male. Sex, therefore, is basically determined by which of the father's sperm penetrates and fertilizes the egg. A normal ejaculate contains as many as hundreds of millions of sperm. Half are X-bearing and half are Y-bearing. They all appear alike through a conventional microscope; only with the use of special techniques can differences be revealed.

One of the holy grails of reproductive biology has been to find the difference between the two types of sperm. Many nonscientific approaches have been developed that claim to capitalize on differences between the X and Y sperm. Some "low-tech" methods have been based on the idea that the vaginal environment may be more hostile to X- or Y-bearing sperm. In this scenario, time of intercourse during the cycle becomes critical, since the quality and biochemistry of cervical mucous (which sperm must traverse in order to enter the uterus) change over the course of the month. One popular technique, practiced in Japan, involves careful charting of the ovulation cycle and use of pH-altering jelly to select for girls or boys. The jelly comes in pink (for girls) or green (for boys) and costs about $100 per vial.[28]

Another nonscientific method, practiced in Canada, involves the belief that Y-chromosome-containing sperm are positively charged and X-bearing sperm are negatively charged. For $499 a company will chart days favorable for conception of a boy versus a girl, based on this notion. (The prestigious journal *Lancet* refused to publish a paper based on this method.)

In fact, X- and Y-bearing sperm *are* different, since the X-chromosome is larger and has more DNA than the Y-chromosome. Since each sex chromosome is only one of twenty-three in sperm, the difference in size and weight of the sperm is not dramatic. The sperm look alike and, until a few decades ago, the difference in X- and Y-bearing sperm could not be measured or documented. But today that difference is measurable. In humans, X-bearing sperm contain 2.8 percent more DNA than Y-chromosome-bearing sperm. That slight but significant difference serves as the basis for the two most promising techniques for preconception sex selection. (In animals the difference in DNA between X- and Y-bearing sperm is greater—approximately 4 percent in cattle and horses. This has permitted the development of a burgeoning industry of sex selection in cows and thoroughbred horses.)

The Ericsson technique, developed in 1973, takes advantage of the differential swimming capacities of X- and Y- bearing sperm.[29] It is reported that sperm that are filtered through a salt-albumin gradient can be separated, since Y-bearing sperm will swim faster into a viscous (thick) solution. Based on a modification of the Ericsson technique, the Midwest Fertility and Sex Selection Center claims a selection rate of 83 percent males for couples who prefer to have boys, and 78 percent females for couples choosing girls. The sperm are processed through a series of protein solutions to yield an enriched population of Y-sperm (for boys) or X-sperm (for girls). The woman is then inseminated with the enriched population requested by the couple. (The cost for this procedure is $500 for the first cycle, $400 for the second, and $300 for each subsequent cycle.)

Many scientists are skeptical about this technique, and there have been numerous publications refuting the results. Research by at least five different groups, using techniques based on these principles, have failed to produce Y-enriched sperm.[30] In fact, another study found the opposite occurring—that the "male oriented samples had a lower Y/ X ratio than female oriented samples."[31] Rose and Wong do note that the separation of Y-sperm can "bias the number of babies born in favour of males," perhaps, they suggest, by inactivating X-sperm. The take-home lesson in this case is that this is still a controversial approach. Clearly the best advice in such an experimental process is "let the buyer beware."

MicroSort—a truly unique and scientifically supported approach to selection of X- and Y-enriched sperm—was developed by Dr. Edward Fugger, Dr. Joseph Schulman, and their colleagues at the Genetics and IVF (GIVF) Institute in Fairfax, Virginia. The technique, which is now in clinical trial, was first announced in a September 1998 article. The article, published in *Human Reproduction*, reported that the technique succeeded in producing the "world's first deliveries of normal babies after use of flow cytometric separated human sperm cells for preconception gender selection. Offspring were of the desired female gender in 92.9 percent of the pregnancies."[32]

The new method, which uses Fluorescent Activated Cell Sorting (FACS), can separate cells of different shapes, weights, or sizes. It takes advantage of the fact that, as discussed above, sperm with an X-chromosome have about 2.8 percent more DNA than Y-containing sperm, since the X-chromosome is larger than the Y. Although this should work well in theory, in fact this method is less than perfect, since there are many sperm of unusual shapes and sizes in every man's semen that confound the process.

As part of the process, the sperm are treated with a fluorescent dye that binds to the DNA. Sperm so treated are exposed to ultraviolet radiation and light up with a different intensity, depending on whether they carry an X- or Y-chromosome. The brighter ones, presumably X-sperm, are shunted into one test tube. The cells that fluoresce less brightly are presumably Y-sperm and are shunted into another test tube. The procedure is very slow, since the sperm have to pass single-file past the FACS detector. Thus, it takes many hours to process one man's sperm sample—resulting in only a small portion of the sperm actually being designated X-enriched or Y-enriched for use in fertilization.[33]

Since there are so few "enriched" sperm, the sperm have to be placed closer to the egg to achieve fertilization. (When natural conception occurs, although millions of sperm are deposited in the vagina, most are lost by attrition as sperm travel through the cervix, uterus, and into the Fallopian tubes.) For most patients, there are enough sperm in the sorted sample that placing the sperm in the uterus (intrauterine insemination, or IUI) can be effective. Some samples, however, have very low sperm numbers and in vitro fertilization (IVF) or intracytoplasmic sperm injection (ICSI) must be used. The cost of sperm sorting plus IUI is $3,200 for one cycle. (It is possible to reduce the cost; by using the ovulation predictor kit at home to chart your ovulation on your own, you can save $900.) The cost of the same procedure together with IVF is about $15,000, and if ICSI is required the cost rises to about $18,000 per cycle.[34]

Rakha Matken, a nurse who manages the daily clinical operations of MicroSort, claims that with "XSort," the current methodology can produce X-enriched sperm that is a mixture of about 90 percent X- and 10 percent Y-sperm. This X-enriched sperm can be used to increase the chances of conceiving a girl from the usual 1:1 to about 9:1. When MicroSort is used to produce Y-enriched sperm, it is less successful. Current statistics claim that "YSort" produces Y-enriched sperm containing 73 percent Y- and 27 percent X-bearing sperm (although only 71 percent of those who wanted boys actually had them). As of this writing, more than 156 babies have been born and more than 317 pregnancies have been achieved using MicroSort. Although a paper was published on the XSort data, it is important to note that the group's data on sex selection for boys have not yet been published.[35]

The authors of the paper on XSort claim that the dye used to tag the sperm has been shown to be safe, since several hundred "normal" research animals have been born using the technique, and to date the miscarriage and birth defect rate does not differ from that expected in the general population.[36] However, other published papers have raised the possibility of damage to the DNA from the binding of the dye. The GIVF Institute is conducting clinical trials now to determine the safety of the technique, and all patients enrolled in its program will be participants in these clinical trials. Although the technique looks promising, it is still not possible to rule out any risk of birth defects resulting from inadvertent DNA damage, since so few human babies have been born from sperm treated in this manner.

One additional method for sex selection that has been used successfully is Preimplantation Genetic Diagnosis (PGD; see chapter 7). In this procedure, the couple must participate in IVF, and all embryos that are produced are allowed to develop to the eight-cell stage. One cell is sucked off of the embryo and tested for the presence of sex chromosomes. If it is XX, the embryo is female. If it is XY, the embryo is male. Since this procedure is extremely expensive, complex, and still experimental, this technique has been performed mainly for couples at risk for conceiving children with serious sex-linked diseases.[37]

Most sex-linked diseases occur when a defective gene is located on the X-chromosome. For instance, the hemophilia gene occurs on the X-chromosome. In females, there are two X-chromosomes, thus, the X-chromosome carrying the normal gene can compensate for the other X-chromosome's hemophilia gene. In males who have a hemophilia gene on the X-chromosome, there is no corresponding normal gene on the Y-chromosome. So a male who inherits the hemophilia gene from his mother (on the X-chromosome, since he gets the Y from his father) will have the disease. Females rarely suffer from sex-linked diseases, since they must get a copy of the disease gene on the X-chromosome inherited from both parents. (And in that case, the father will also have the disease, as his only X-chromosome carries the disease gene.)

By selecting for the sex of the embryo, the couple can choose to have only girls and avoid having a child with hemophilia, Duchenne muscular dystrophy, or X-linked hydrocephalus, for example. PGD allows this selection to occur before pregnancy is established, hence it is permissible according to many rabbinic authorities.

The Slippery Slope of Sex Selection

In September 2001, a startling shift in policy was announced by John Robertson, acting chairman of the American Society for Reproductive Medicine (ASRM) ethics committee. Robertson wrote that in vitro fertilization, preimplantation genetic diagnosis, and embryo sex selection could be used to generate gender variety in a family. In other words, a family with one or more boys could use IVF and PGD to ensure that the next child is a girl. A family with one or more girls could guarantee that the next birth would be a boy. Using this procedure, embryos of the unwanted gender would not be chosen for implantation.

A previous policy statement by the ASRM had discouraged use of embryo sex selection simply to satisfy a gender preference of parents. But Dr. Norbert Gleicher of the Center for Human Reproduction, an organization that has nine fertility clinics in the Chicago and New York areas, had asked the ASRM for a policy statement on this application. His clinics plan to apply the technique of embryo sex selection solely to enable parents to choose what gender they want. Gleicher reasoned that if sperm sorting is acceptable for gender selection, why not use PGD, a "superior method"? PGD used for embryo sex selection can virtually guarantee that an embryo of a particular sex is implanted.[38]

Many reproductive experts cringed at the prospect of such an application. Dr. William Schoolcraft of the Colorado Center for Reproductive Medicine challenged, "As we learn more about genetics, do we reject kids who do not have superior intelligence or who don't have the right color hair or eyes?" He explained how he viewed the difference between sperm sorting and embryo selection: "With sperm sorting, you are not throwing away potential babies."[39]

Dr. James Grifo, a reproductive endocrinologist at New York University Medical Center and president of the Society for Reproductive Technology declared, "Sex selection is sex discrimination, and I don't think that is ethical." Grifo also revealed that the demand for this technique could be so great that "I could have financed all my research from now until the day I die if I honored all the requests."[40]

Although some clinics plan immediately to go ahead with this application, it remains to be seen whether the backlash will result in legislation against using embryo selection for gender selection, especially with no medical justification. Just as with the use of selective abortion in China, India, and Korea, the ability to choose, with certainty, the sex of an embryo, could lead to skewed sex ratios. (Of course, the high cost of IVF, PGD, and embryo selection would limit its use only to people with enough money. Poor people would not have the option of using this "foolproof" method to select the gender of their children.)

Guidelines for Sex Selection

Clearly, wholesale use of sex selection could affect birth ratios, and the resultant skewing of ratios could seriously impact society. Thus, religious and secular ethicists have developed guidelines for the application of these techniques.

Liu and Rose have stipulated the following rules for application of sex-selection technology (this list was generated for use with technologies such as sperm sorting): That couples be married or at least in a stable relationship; that no childless couples be accepted; that the procedure be done only to achieve family balance; that the couple should submit proof of the first three qualifications (their committed relationship, their preexisting child(ren) and the gender(s) of those offspring); that the couple signs a contract agreeing not to abort any child of the "wrong sex" who may be conceived via these techniques; and that the couple understands that there is no 100 percent guarantee of success in choosing a sex.[41]

The Pennings guidelines also stipulate that sex selection not be used for the first child in the family or in families that are already balanced.

The sex selected for must be the one that is less represented in the family, and the only exception can be when there is a risk for a genetic disease in one sex (e.g., an X-linked disease).[42]

The guidelines on sex selection for the GIVF Institute, when used for genetic disease prevention, include the following: "Married couple; Known carrier of sex-linked or sex-limited disorder. . . ." However, when used for family balancing, the required qualifications are: "Married couple; Must have at least one child; Sort for the less represented sex of children in the family. . . ."[43]

Rabbinic Reactions

"A man is required to sire both a son and a daughter in order to conform with the biblical mandate 'be fruitful and multiply,'" states Rabbi J. David Bleich. Many observant Jewish couples have large families not only to fulfill the one son/one daughter mandate, but also because certain methods of birth control are not acceptable by strict halakhic standards. (Exceptions are made if pregnancy would jeopardize the health of the mother.) Must a couple resort to unusual lengths to fulfill the one son/one daughter mandate? "A couple who are the parents of any number of children of one sex are under no obligation to utilize any artificial method in order to increase the chance of the birth of a child of the other sex," affirms Bleich. "The birth of a son and a daughter is simply the point at which there is no longer a pentateuchal obligation . . . to continue to perform acts which make procreation possible."[44]

The fact of the matter is, however, that Jews may be interested in sex selection of offspring for a variety of reasons. Would such actions be permissible?

Rabbi Elliot Dorff adheres to the principle that there should be a compelling health-related reason to screen and intervene in a pregnancy. According to Dorff, "It is generally *not* permissible, according to all interpreters of Jewish law, to screen specifically for gender just because one wants a boy or a girl or to screen for any characteristic other than disease (e.g., height or intelligence). Screening for gender is thus only acceptable when there is a family history of gender-related disease linked to the chromosome for the child's gender. (Most such diseases affect males.)"[45]

Some rabbinic leaders are even more troubled by the prospects of the technique. "On a philosophical level, I see a vast difference between a couple that is motivated by concern for a genetic disease *versus* a couple who is motivated by a family planning concern," considered Rabbi Shmuel Goldin. "I think . . . we have to be very, very careful about over planning our populations, over planning our families. There's a certain degree of God-given, from our perspective, 'serendipity'—and to try to usurp that is very problematic."[46]

According to Dr. Richard Grazi, there is unanimity of Orthodox rabbinic rejection regarding sex selection. He reports that Rabbi Yigal B. Shafran, Director of the Department of Halakhah and Medicine of the Jerusalem Religious Council, rejects sex selection. Similarly, Rabbi Yitzhak Zilberstein of the Israeli Medical-Halakha Group forbids it. "Rabbis Shafran and Zilberstein regard the rabbinic judgement to be that securing a baby of a desired sex is simply too frivolous a halakhic concern—no matter how pressing it might be for the specific couple— to overcome the other moral objections," writes Grazi.[47]

Rabbi J. David Bleich states unequivocally that "no authority would accept sex determination as legitimate cause for abortion." He also expresses concern that sex selection will result in "a marked increase in male over female births." Bleich admits that "A yearning for male babies by prospective parents appears to be manifest in virtually every culture. This is true not only of prospective fathers, but of prospective mothers as well." He cites several studies from the 1970s supporting this statement, which indicate that "at least two-thirds of American women express a preference for sons rather than daughters."[48] It should be noted, however, that more recent studies of American couples who consult with reproductive clinics do not indicate a strong preference for boys over girls.

Hashkhatat hazera, or destruction of the seed, is also of concern to Bleich. The procedure used in both the Ericsson and MicroSort techniques involves separation of X- and Y-bearing sperm and disposal of the unwanted portions. "Separation of the androsperm (for males) from the gynosperm (for females) . . . [involving] destruction of either the male-producing or the female-producing sperm" would constitute a violation of *hashkhatat hazera*. "Hence an attempt to determine sex in this manner would be a violation of Jewish law," concludes Bleich.[49]

Rabbi David Feldman, an expert in Jewish medical ethics, also voiced concern about the power to choose a baby's sex. "The human ecology would be sent askew, because the randomness of the number of males and the number of females helps the world as it is. In China they select the gender of the child. . . . [N]ow they don't have enough women for men to marry."[50]

"It can also affect the total population," Feldman continued. "Sometimes a couple has six children because they are waiting for a child of the opposite sex. . . . [T]hey might have six sons or six daughters. If they could have selected the gender they might have stopped at two, one of each. The halacha is that the *mitzvah* of *p'ru urvu* [be fruitful and multiply] is fulfilled with one of each, a son and a daughter. They might say they've fulfilled the *mitzvah* and they don't have to go further."[51]

On the use of the technique to avoid serious illness, Feldman was more positive. "I would be supportive because it is a health matter, it's not a whimsical or convenience matter," he asserted. "Gender selection to prevent the birth of hemophiliacs is a more substantive reason, but even that may not be enough of a reason, depending on the therapeutic picture. But of all the reasons it's the best."[52]

There are important halakhic issues regarding the use of IVF, PGD, and embryo sex selection. On the one hand, choosing specific embryos to use and others to discard may be acceptable according to some rabbinic sources. However, with regard to any medical procedure, *pikuakh nefesh* (preservation of life) is of primary concern. In vitro fertilization is a fairly safe and routine technique, but in some cases it could pose hazards for a woman. The hormonal treatments used to induce superovulation can result in the growth of ovarian cysts and in other complications. Egg recovery methods are invasive and carry risks of infection, hemorrhaging, and even serious damage to the female reproductive organs. "Since in vitro fertilization is not free of complications, I do not believe women should undergo IVF for the sole purpose of sex selection," said Dr. Zev Rosenwaks, Director of the Center for Reproductive Medicine and Infertility at Cornell University.[53] Thus, this is not an approach to be chosen lightly or for frivolous reasons by anyone.

One Jewish Mother

"I always envisioned girls because I'm from a two girl family, so I just imagined families having girls, and I felt miscast as the mother of boys," confessed Ellen,[54] an Orthodox Jewish mother of three boys. "Of course I love them dearly, but there were a few years there at the beginning . . . when I really thought that I was totally miscast for this role and that 'up there' they made a mistake. . . ."

Would Ellen have used a sex-selection technique, if it were available, to conceive a girl? "Your eye sees and then your heart wants. If your eye doesn't even know it's available, your heart can't start to want. So I guess I never really allowed myself to start thinking," she admitted.

Clearly, many parents do give much thought to the matter. Couples who have girl after girl may not express it out loud, but might, if they could, push the odds in favor of a boy. And in this country, at least, it works both ways, as attested to by Ellen, a happy and well-adjusted mother of three boys, who had always imagined herself as a mother to daughters as well.

Sex selection would be used by most couples for "family balancing," explained Dr. Lee Silver, Princeton University professor and author of *Remaking Eden,* a book about human cloning. "For Americans the perfect family is one boy and one girl. . . . If used for family balancing . . . if they have two boys, they want a girl, if they have two girls, they want a boy"; with such an application he feels there should be no problem—in the United States we would end up with a balanced population.[55]

"The big worry that I would see, and I don't think it's likely in the U.S. but certainly in other parts of the world," Dr. Ruth Macklin, Professor of Bioethics at Albert Einstein College of Medicine, conceded, "is an imbalance in the sex ratio. There's already an imbalance in countries such as China and India . . . because there is such a preference for males rather than females. There are many more men and too few women."[56]

Prudent use of this new technique would take into account its possible risks, its brief track record and the philosophical concerns of clergy and ethicists. "It's one of the areas in which a choice is not a good idea," concluded Feldman. "Too many choices are not necessarily a blessing. . . . There's so much we don't know about this exquisite balance and what it takes to preserve it."[57]

On a practical level, most couples will still opt for the mystery of natural conception. "We just make ourselves happy . . . because it's out of our hands," mused Ellen, explaining her perspective on family planning. "I have already learned how to be very happy with what I have. . . . I just want whatever *Hashem* [God] wants me to have."

CHAPTER NINE

TAG A CAT:
Jewish Genes and Genealogy

I REMEMBER HEARING as a child, that only Kohanim, or Jews of Priestly descent, were physically able to make the special gesture for priestly blessing—to split the fingers between the middle and ring finger (similar to the Vulcan greeting of Star Trek fame). Priestly traditions are transmitted from father to son in a line that stretches back 3,500 years to Aaron the High Priest. Since Priestly status is transmitted from father to son, there should be a common genetic heritage of all Kohanim. Indeed, the estimated 350,000 Kohanim around the world are all closely related descendants of Aaron, and, in fact, DNA studies have confirmed that in addition to their religious and cultural links, there are biological links among this group. If this is the case, could we find one day on the drugstore shelf a confidential do-it-yourself test for Jewish Priestly DNA? Before we answer that question, we must first discuss what DNA is and how it determines human identity.

DNA, and the concept of a genetic code, has always excited me. Until recently I thought of DNA as kind of a black hole, or dark continent, into which we could peek but not really explore. Now that the Human Genome Project has succeeded in decoding an initial draft of the genetic code, scientists can make in-depth explorations to reveal the details of the instructions for building human cells and human beings. The instructions are written in four chemical building blocks, called the nucleotide bases. These bases, whose full names are abbreviated A, C, G, and T, are strung together in a specific order to direct the cellular machinery to build structures and create working units—the living cells of the body. It takes approximately 3.2 billion DNA bases (the As, Cs, Gs, and Ts of the DNA alphabet) to encode the instructions for a human being. And the details of these instructions are so specific that a single change or mutation of one DNA letter can spell disaster for the cells and, indeed, the organism. Scientists are busy deciphering the billions of letters; they are trying to make sense of the complex code. These DNA scholars—men and women who will devote decades to the study of the DNA text—will help to demystify life.

TAG A CAT is an example of a hypothetical genetic sequence. The seven letters chosen here happen to spell out a brief English sentence. I composed this short sentence (which describes a playful game) to illustrate a point: That any change in the letters of this sentence will change its meaning. For instance, if you change the C to a T, the sentence becomes "TAG A TAT, which has no meaning in English; it is nonsense. If you change the first T to a G, it becomes "GAG A CAT," which has an entirely different (and somewhat macabre) meaning. Likewise every detail of genetic sequence, every DNA base, can be critical. And even the slightest alteration in the DNA could change the meaning, or even destroy the activity of the gene, making it impossible for the cell and organism to function properly. Some of the mutations that occur in DNA sequences are called nonsense or missense mutations—since they destroy the meaning of the code and can harm the cell and organism.

It turns out that of the 3.2 billion DNA letters, only about 1.4 percent actually direct the cell to make proteins.[1] That means there are approximately 45 million DNA letters that provide genetic information to guide the cell to build proteins, the fabric of the cells. The DNA letters, or bases, make up the genes, and the genes specify, in a stepwise fashion, like a blueprint, how the cellular machinery should construct proteins. And proteins take care of the rest: They work as catalysts to drive chemical reactions; they make up structures, such as the cellular scaffolding that determines the shape of cells; they determine how cells communicate with each other to live in harmony and produce a coordinated living being; and they ensure that the organism will develop and proceed through all of life's stages, as programmed.

This 1.4 percent of the DNA code holds the instructions for about 40,000 different genes. Before the Genome Project succeeded in decoding the human genome, scientists assumed that it would take at least 80–100,000 different genes to build a human being. The lower estimate of 40,000 was truly a surprise, especially since humans turned out to be not much more complex than some lower species. For instance, we have only five times as many genes as baker's yeast, a simple single-celled creature. We have about twice as many genes as the lowly fruit fly or a simple worm. This information should bring us down to earth and enable us to better understand our beginnings. It is sobering to realize the small amount of genetic information that differentiates us from the lower creatures.

Francis Collins, director of the National Human Genome Institute, has described the genome in this way: "We've called the human genome the book of life, but it's really three books. It's a history book. It's a shop manual and parts list. And it's a textbook of medicine more profoundly detailed than ever."[2] The human genome can teach us about various aspects of the Jewish people. As a history book, it can reveal relationships between populations and can uncover patterns of migration. As a shop manual, parts list, and textbook of medicine, it can help us understand how things go wrong, and it can explain disease, especially the diseases that are prevalent in the Jewish community.

Judaism is a religion, not a race. That is why we see a great diversity among the Jewish people: The Jewish community includes blonds, brunettes, and redheads, swarthy Mediterranean types, fair northern Europeans, black Ethiopians, Jews from India with Asian features, and many other forms and phenotypes. That said, it is important to note that the Jewish people, although mixing with local groups and assimilating throughout history, retained its identity by discouraging conversion, by minimizing contact with outside groups, and by maintaining a more or less cloistered existence throughout thousands of years of wandering. And the "history books," which are the human genomes of many Jews, contain remnants of common origins and descent from common ancestors. That is what makes a study of Jewish genes so compelling. Many Jews are linked by common genetic threads, which can manifest as "Priestly genes," or Ashkenazic or Sephardic recessive disease genes, or cancer-causing BRCA genes (see chapter 10). All of these "Jewish genes" are genetic legacies of a tightly knit group.

The Y-Chromosome: Father's Legacy

It is not possible to compare *all* the genes of two human beings—that would be a daunting task. Thus, when scientists want to compare genetic heritage, they choose a subset of genes to study. When selected genes are studied, it is still possible to get a good idea of how individuals and populations are related to each other. The 3.2 billion DNA bases of genetic information in each human cell are arrayed on forty-six chromosomes—their genetic packaging—and they reside in the nucleus, a large compartment in each cell. Although genetic studies can be done using genes from any of the forty-six chromosomes, some of the most useful studies have been performed on DNA sequences located on the Y-chromosome genes. There are several reasons for this.

First, when sperm and eggs are produced, all of the chromosomes go through a process called crossing over, where chromosome pairs (which each person originally inherited from his or her own mother and father) line up and exchange genetic sequences. This process occurs in a random fashion, resulting in new arrangements of genetic combinations in the sperm or egg, which can then be passed on to the next generation. The exception to this rule is the Y-chromosome. Since most of the Y-chromosome has no equivalent region on the X-chromosome, it rarely swaps segments with the X. So, generation after generation, from father to son to grandson, the Y can remain more or less unchanged.

Second, although the Y-chromosome does not swap major segments with other chromosomes, it does change over time. DNA is not static; it undergoes spontaneous changes, or mutations. Those changes happen if errors occur as the DNA is copied (replicated) before cell division. Mutations can also occur as a result of environmental damage resulting from chemicals and/or radiation. Mutations occur in a random fashion and can accumulate over a period of time. Because we can estimate the rate at which random mutations will appear in DNA, by studying the differences in a specific DNA sequence in two populations with a common ancestor, it is possible to estimate how long these two groups have been separated from each other. The longer the groups have been apart, the more differences there will be between their DNA codes.

The third reason that Y-chromosome studies are useful, is that genealogy can frequently be traced more readily through the male line. In male-dominated societies, patrilineal inheritance of property and surnames is common. A notable exception is the matrilineal transmission of Jewish status (according to Orthodox and Conservative Jewish views). Although matrilineal descent is important for determining Jewish status, within the ancient Israelite community tribal membership was determined by the male. And nowadays, according to traditional Jewish groups, membership in the Israelite, Levite, or Priestly caste is still determined by patrilineal transmission.

Mitochondrial DNA: Mother's Legacy

There is a chromosome in human cells that is inherited maternally—that is, through the egg, not the sperm—and it has also been used successfully to track maternal lineage. It is an unusual chromosome, since it is not one of the forty-six and is not located in the nucleus of the cell. This unique piece of DNA is called mitochondrial DNA, or mtDNA. A miniature loop of DNA, mtDNA is found in another cellular compartment, the mitochondrion rather than the nucleus.

This small structure (which provides energy for the cell) carries DNA equivalent to about 1 percent of the cell's total genetic material. The genes carried by mitochondria are unique; they do not appear on the other forty-six chromosomes. All cells have mitochondria, including sperm and eggs; but when fertilization occurs, only the mitochondria of the egg remain in the embryo. So only the mother's mitochondria are passed on to the next generation. Studies of mtDNA from human subjects all over the world have permitted researchers to propose a geographic region for the origin of the human species. The DNA evidence based on mtDNA studies suggests that our common ancestor is a woman, whom scientists have dubbed the "mitochondrial Eve," who probably lived some 200,000 years ago in Africa.[3]

There have been landmark discoveries in Jewish genealogy based on Y-chromosome genes, and advances have also been made based on inheritance of disease genes located on other chromosomes (see chapters 6 and 10). These types of studies have permitted a glimpse into understanding the inheritance of "Jewish genes." To date there is very little information on mtDNA with regard to Jewish populations;

however, we can anticipate discoveries on maternal lineage, as a number of labs appear interested in studying Jewish genealogy through the maternal line.

Do Priestly Genes Exist?

In ancient Israel, the Priests and Levites were required to serve in the Temple in varied capacities. They conducted special services during the festivals, they tended to the sacrifices brought by Israelites, and they maintained the Temple and its holy vessels. The role of the Israelites was to worship and bring sacrifices and tithes to the Temple, in honor of important life-cycle occasions and on the three major festivals of the year. After King Solomon's reign ended, Israel was split into two kingdoms. The northern segment, the Israelite Kingdom, was destroyed in 722 B.C.E., and ten of the original tribes were exiled, never to return to the Holy Land. (As we will discuss later, there is evidence that many of those groups did not disappear; rather they stayed together in remote areas and continued to practice some variant of the Israelite religion. Some of their descendants have been discovered in parts of Asia, Europe, and Africa.)

At the time of the Israelite exile, the southern Kingdom of Judea—including the tribes of Judah, Benjamin, the Levites, and the Priests, remained undisturbed. But in 586 B.C.E. they, too, were conquered, the Temple was destroyed, and they were exiled to Babylonia. Those Judeans did eventually return from exile and rebuilt the Temple, permitting the Levites and the Temple Priests to maintain their distinctive identities via Temple service. After the Second Temple was destroyed in 70 C.E., and the Jews were again exiled, the Levites and the Priests were assigned new duties in synagogue service. Benjamin and Judah, the remaining Israelite tribes, were also dispersed, and although they still identified themselves as Israelites, people lost track of their specific tribal origins. Thus for almost two thousand years, tribal status for most Jews has meant being classified as Israelite (lowest caste), Levite, and Kohen (Priest—highest status). Membership in each group has been passed from father to son for about eighty generations since the last exile. In synagogue service, the Kohanim are called first to the Torah, Levites are second, and Israelites are last. And at daily prayers (in Israel) and festivals (all over the world) the Kohanim bless the congregation (using the distinctive split-finger gesture). Since people who convert into Judaism are assigned Israelite status, the Kohanim and Levites of today should, theoretically, be able to claim a purity of lineage—unmixed with other groups—and entirely patrilineal in nature.

The purity of this lineage was critically important, since Temple service required incontrovertible Priestly lineage. Many chapters in the Mishnah are concerned with a person's status as Priest or Israelite. One particular passage dealing with the mix-up of two infants shows how important this status was:

> If the newly-born child of a priest's wife were confused with the newly-born child of her bondwoman, then both may eat of priest's-due and they receive their share together at the threshing-floor and they must not contract uncleanness for the dead and they must not marry women whether they are eligible or whether they are ineligible. If they grew up still confused and then freed each other, they wed women eligible for priestly family and they must not contract uncleanness for the dead . . . and the strict rulings regarding priests and the strict rulings concerning Israelites apply to each one.[4]

Taboos and rules of conduct regarding Priests have been carefully adhered to by religious Jews throughout the ages. Priests must maintain ritual purity—they must not come into contact with dead bodies, they may not enter cemeteries, and they may not marry divorcees.

If the integrity of the lineage has been preserved, the Kohanim and the Levites should carry a distinctive Y-chromosome, passed down from father to son over 140 generations from the original Priestly family headed by Aaron, brother of Moses. The differences between Y-chromosomes from one Kohen to another should only be a function of those occasional random mutations. Israelites should have more heterogeneous Y-chromosomes, since they descended from a more heterogeneous population, including males who converted to Judaism.

Karl Skorecki, a Kohen and a research scientist at Rambam-Technion Medical Centre in Haifa, Israel, had first wondered about the common genetic heritage of Kohanim when he was worshiping in synagogue and observed a Sephardic Kohen of Moroccan descent being called to the Torah. (As mentioned above, Kohanim are afforded special privileges during synagogue services, and when there are Kohanim present in synagogue they are called first to make a blessing on the Torah.) Despite differences in skin, hair, and eye coloration observed by Skorecki, he noted that both he and the Moroccan

Kohen shared the Priestly heritage and traditions, and he wondered what else they might share from a genetic standpoint.[5] Skorecki organized a research team of scientists from Israel, Canada, England, and the United States to perform a study of DNA sequences, comparing DNA from Kohanim with DNA from other Jews who identify themselves as "Israelites." (Levites, or descendants of the special tribe of Levi, were omitted from the "Israelite" group.)

Cell samples were obtained from 188 Jewish men from Israel, North America, and England. Sixty-eight of the men identified themselves as Kohanim, and the rest identified themselves as Israelites. Special DNA sequences, unique to the Y-chromosome, were studied. Each Y-chromosome has unique sequences, called markers, that can appear in different forms in different individuals. One marker was found in only 1.5 percent of the Kohanim, as compared to 18.4 percent of the Israelites. Another marker appeared in 54 percent of the Kohanim, but only 33 percent of the Israelites. These findings suggested that a large portion of the Kohen population carries common genes inherited directly, through patrilineal descent, from one male ancestor. In other words, many priests carry the Y-chromosome, which presumably originated with Aaron the High Priest 3,500 years ago. Results of DNA tests were similar in Ashkenazic and Sephardic Jews, which further supports the universality of the genetic findings.[6]

In a follow-up paper, Skorecki and his colleagues characterized Y-chromosome DNA sequences called haplotypes. A haplotype is a series of known DNA sequences that occur together (or are linked) on the same chromosome. They discovered that there was extensive diversity on Y-chromosomes of Israelites (i.e., there are many different haplotypes in the Israelite population). But a particular group of six chromosomal markers—that is, a single haplotype out of the 112 different haplotypes studied—turned up in high frequency on Y-chromosomes of Ashkenazic and Sephardic Kohanim. More than 91 percent of Kohanim (97/106) exhibited one such cluster of genetic markers, whereas only 43 percent (35/81) of Levites and 62 percent (74/119) of Israelites had those same markers. A more comprehensive group of twelve distinct genetic markers, defined as "the Cohen Modal Haplotype" (CMH), was used to further define Y-chromosomes in genealogical groups. Since both Ashkenazic and Sephardic Kohanim had CMH in very high frequency (44.9 percent of Ashkenazic Kohanim and 56.1 percent of Sephardic Kohanim, compared with 13.2 percent of Ashkenazic Israelites and 9.8 percent of Sephardic Israelites), this demonstrates that despite the fact that Ashkenazim and Sephardim have been isolated from each other for over 500 years, both Priestly communities have a common origin, and the "Cohen chromosomes were derived from a common ancestral chromosome."[7]

As opposed to the Kohanim, many of whom have common genetic markers, it turns out that Levite Y-chromosomes are surprisingly diverse. Historically, the Levites are also said to have originated from a common ancestor—Levi, the son of Jacob, founder of the tribe. However, the 1998 study suggests that "contemporary Levites . . . are not direct patrilineal descendants of a paternally related tribal group."[8] How could this have happened? Perhaps over the generations there were individuals who decided to "pass" as Levites so as to increase their status in the community. People motivated to sneak into the "Levite club" may be more reluctant, or even afraid, to cross the next boundary into the Priestly caste.

The data from CMH analysis also permitted the authors to estimate how long ago the common Priestly ancestor lived. This type of calculation is based on the differences found among the modern chromosomes. By calculating the expected rate of mutation per generation, they estimated that it has been some 106 generations since the common ancestor lived. If average generation time is estimated at 25–30 years, then the common ancestor lived approximately 2,650–3,180 years ago (which falls between the time of the Exodus from Egypt and the destruction of the First Temple). That is consistent with the time of the First Temple period and Aaron and his sons and grandsons (plus or minus a few hundred years!). Thomas and coworkers concluded that "although Levite Y chromosomes are diverse, Cohen chromosomes are homogeneous. We trace the origin of Cohen chromosomes to about 3,000 years before present, early during the Temple period."[9]

Are the Lemba Jewish Tribesmen?

Y-chromosome analysis, which revealed the presence of unique genetic sequences carried in high frequency by Kohanim, has also been used to determine genetic links to other groups claiming Jewish ancestry. The most interesting example of this type of analysis has been regarding the Lemba. The Lemba are a group of people in southern Africa who claim to be descended from Israelite males who migrated by boat from a place called "Sena in the north." Different sources have suggested the location of ancient Sena as being in Yemen, Judea, Egypt, or Ethiopia. Tudor Parfitt has claimed to have evidence that the site of ancient Sena was in Yemen.[10]

While most Lemba today practice forms of Christianity or Islam, they retain some practices that harken back to ancient Judaism.[11] For instance, they circumcise their sons, although instead of doing it in infancy, which is the Jewish practice, they perform a circumcision ceremony at age fourteen. They also adhere to strict dietary laws reminiscent of Judaism, namely, no pork is permitted and they do not mix meat with milk.[12]

The Lemba are divided into more than twelve separate clans, the most important of which is the Buba clan. Legend has it that a leader by the name of Buba led the Lemba out of Judea; their travels eventually landed them in southern Africa—where they now reside in Mozambique, Botswana, and South Africa.[13]

Thomas and collaborators (including Tudor Parfitt, Karl Skorecki, Neil Bradman, and David Goldstein, the scientists involved in the first CMH research) studied DNA sequences from Lemba, Bantu, Yemeni, Sephardic Jews, and Ashkenazic Jews. DNA analysis of their Y-chromosome haplotypes revealed that there have been African Bantu and Semitic contributions to modern Lemba genes. However, "Support for a Jewish contribution to the Lemba gene pool is, nevertheless, found in the presence, at high frequency in the Lemba, of the CMH." One of the Lemba clans, the Buba, exhibited a very high frequency of CMH, "which is known to be characteristic of the paternally inherited Jewish priesthood and is thought, more generally, to be a potential signature haplotype of Judaic origin." In other words, groups that show a high level of CMH in their Y-chromosomes are thought to have Jewish ancestry. This conclusion can be supported, since the frequency of CMH in the Jewish priesthood is estimated at about 50 percent; CMH appears in about 12 percent of lay Jews; and CMH is undetectable or found only at low frequencies in a variety of non-Jewish populations, such as Greeks, Armenians, Mongolians, Yakut, Nepalese, and Cypriots.[14]

Why would lay Jews (i.e., Israelites) carry the CMH at all? Perhaps gene flow from the Priestly to the non-Priestly castes occurred. Koha-

nim could have abandoned the fold upon marrying divorcees or violating other Priestly taboos. Some families might simply have lost
their thread of tradition by forgetting the customs and responsibilities, or by making a conscious decision not to transmit it to their
sons. The amazing thing is that so many Kohanim did transmit the
customs from generation to generation, throughout thousands of
years of exile, despite having no Temple in which to serve.

Neutral and Survival Genes

Genetic markers that have been studied to determine ancestry and
genetic relationships include two major categories: neutral markers (genetic sequences that should neither help nor hinder survival), and genetic markers that confer some selective advantage on an individual.
For instance, fingerprint patterns (arches vs. whorls vs. loops) are believed to be neutral markers. The pattern on your fingertips will not
influence your survival in a desert climate, a high elevation, or a
swamp. (It should be noted, however, that the genes determining fingerprint patterns may be closely linked to genes for other traits. Any
increased survival value for those other traits may influence the transmission of particular fingerprint patterns.) On the other hand, blood
group markers (A, B, and O, which determine your blood type) may
themselves confer advantages for survival under some circumstances.
The gene for blood group B may confer a resistance to smallpox; in
some regions, this was critical to survival. Those with blood group A
were more susceptible to smallpox and less likely to survive.[15]

The neutral genes tend to be distributed randomly within a group
of people no matter what type of environment they live in. Neutral
genes thus may be passed on, generation after generation, and can
serve as genetic legacies useful for elucidating ancestry and genetic relationships within and among different groups of people. On the
other hand, genes that confer advantages or disadvantages on individuals would be selected for or against in particular environments. So
people who move to a new area with different environmental conditions (for example, climate, diseases, foods, or predators) may, over
time, exhibit changes in frequencies of those genes—and have patterns similar to the natives of that area. People with genes that confer
advantages are more likely to survive, and others are more likely to

die. That is why a population that moves into an area begins to resemble the native population vis-à-vis these types of genes; since the host population and the migrant population would both benefit (or suffer) as a result of certain genes.

The Jewish Diaspora has resulted in populations of Jews living in all corners of the Earth. Neutral genes have tended to stay with the Jewish population wherever they migrate. Y-chromosome markers that have been used in studies of Jewish genealogy appear to be neutral DNA sequences. They do not confer an advantage or disadvantage, hence, as we have seen above, markers such as the Cohen Modal Haplotype can link the ancient and the modern Jews. Likewise, fingerprint patterns of Ashkenazic Jews resemble their ancient Arab and Egyptian neighbors more closely than their more recent European neighbors. In fact, in describing fingerprint patterns, Dr. Jared Diamond reports, "Jews, Arabs, southern Italians and other Mediterranean peoples have lots of whorls, while northern Europeans have many loops and few arches, and Chinese have as many whorls as loops." However, the ABO blood groups of the wandering Jews have evolved—probably because of natural selection, and "these same [Ashkenazic] Jews have by now become thoroughly Germanified in terms of their ABO blood groups."[16]

Of course, a certain amount of conversion and intermarriage should be able to account for the fact that Jews tend to resemble the gentile host population where they reside. But the fact that neutral genes remain unchanged challenges the idea of extensive mixing of the populations. So, as explained above, the reason that Jewish populations resemble their host populations vis-à-vis certain genes must be because of selective pressure that prefers certain traits advantageous to survival. In addition to ABO blood groups, those traits could include skin color and genetic enzyme deficiencies. Indeed, "Ashkenazic Jews . . . tend to be more fair-skinned and more likely to have blond hair and blue eyes than are Sephardic Jews from the Mediterranean. In turn, the Jews of Ethiopia and India are, on the average, darker-skinned than Mediterranean Jews, just as ethnic Ethiopians and Indians are usually darker than Egyptians."[17] This type of natural selection would therefore explain why wandering Jews look like the host population where they reside, but still maintain a Jewish genetic signature, such as the CMH.

Another gene, one that protects against disease, also appears to be selected for in certain environments. G6PD deficiency, a genetic condition that causes anemia, confers protection against malaria. A deficiency in the enzyme G6PD could arise in a population as the result of selective advantage if they live in an area where malaria is a threat. And, predictably, Jews who migrated to Mediterranean regions where malaria was a problem eventually resembled the host populations with regard to G6PD deficiency.[18]

Do Other Groups Have Jewish Origins?

Given the genetic similarities and differences found between Jews and their host populations, and the genetic similarities between Jewish groups who have been separated for dozens of generations, can we genetically define "Who is a Jew?" And if we can, does that allow us to include or exclude groups at will in our definition?

A study by Michael Hammer and colleagues asks, "Given the complex history of migration, can Jews be traced to a single Middle Eastern ancestry, or are present-day Jewish communities more closely related to non-Jewish populations from the same geographic area?" They studied 13 Y-chromosome haplotypes found in Jewish populations to "trace the paternal origins of the Jewish Diaspora." The study included 1,371 male subjects from 29 different populations, including 16 non-Jewish groups and 7 Jewish groups (which included Ashkenazi, Roman, North African, Kurdish, Near Eastern, Yemenite, and Ethiopian).[19] The analysis suggested that only a low level of genetic mixing occurred between Europeans and Ashkenazi and Roman Jewish groups. "Despite their long-term residence in different countries and isolation from one another, most Jewish populations were not significantly different from one another at the genetic level." One surprising finding was that the six Jewish populations with close genetic ties to each other also closely resembled two of the Middle Eastern populations, Palestinian Arabs and Syrians.[20]

In general, the farther apart two populations of people reside, the more different they are genetically. Hammer's report confirmed this general rule with regard to non-Jewish populations studied. (Germans and Italians are geographically *and* genetically close, both groups are more distant geographically *and* genetically from Egyptians, farther removed from Ethiopians, and most distant from Zulus.) However, this was not found to be the case with the Jewish groups studied. The study points out that, although thousands of miles separated African, European, and Middle Eastern Jewish groups, "the level of divergence among Jewish populations was low despite their high degree of geographic dispersion." "In fact," states the report, "these Jewish populations had the lowest ratio of genetic-to-geographic distance of all groups in this study."[21]

The authors conclude that there is a common genetic origin for six of the seven Jewish populations studied. "The results support the hypothesis that the paternal gene pools of Jewish communities from Europe, North Africa, and the Middle East descended from a common Middle Eastern ancestral population, and suggest that most Jewish communities have remained relatively isolated from neighboring non-Jewish communities during and after the Diaspora." Statistical analysis of their data revealed that there was little mixing of Ashkenazi Jewish populations with European non-Jews. "If we assume 80 generations since the founding of the Ashkenazi population, then the rate of admixture would be less than 0.5% per generation."[22]

There was one Jewish group that did not match the other six genetically: the Ethiopian Jews. Ethiopian Jews seem to be the exception to the genetic similarities found in Jewish groups around the world. Thousands of Ethiopian Jews migrated to Israel in a series of 1980s Israeli airlifts (Operation Moses) and a one-day 1991 airlift (Operation Solomon).[23] Also called Falashas or Beta Israel, many Ethiopian Jews claim descent from King Solomon and the Queen of Sheba,[24] but many scholars believe they are descended from a group that converted to Judaism in the fourteenth or fifteenth century.[25]

In the Hammer study, six Jewish populations exhibited high frequencies of two haplotypes, called Med and YAP+4S; however, Ethiopian Jews had high frequencies of two entirely different haplotypes, 1A and 4L. In terms of haplotype frequency, the Ethiopian Jews most closely resembled their non-Jewish Ethiopian neighbors. Interestingly, the Lemba showed some similarities to sub-Saharan African *and* to Jewish populations. Approximately 40 percent of Lemba Y-chromosomes studied resembled the African groups rather than the Jewish ones.[26]

In earlier studies, Ethiopian Jews were also shown to differ significantly from other Jewish populations. Ritte and coworkers studied DNA haplotypes from Y-chromosomes and from mitochondrial DNA. Y-chromosomes, as explained above, are passed from father to son—

hence, they establish a paternal line of inheritance. Mitochondrial DNA, or mtDNA, is inherited through the mother. These investigators determined that, for both paternally and maternally inherited markers, Ethiopian DNA differs significantly from Ashkenazic, North African, Near Eastern, Yemenite, and Minor Asian/Balkanian Jewish groups.[27]

Lucotte and Smets also presented genetic evidence that Ethiopians did not stem from ancient Israelites. Y-chromosome haplotypes were studied, revealing that two haplotypes, which they call haplotypes V and XI, occur most frequently in Falashas and Ethiopians, but the "Jewish haplotypes VII and VIII are not represented in the Falasha populations." The authors concluded "that the Falasha people descended from ancient inhabitants of Ethiopia who converted to Judaism."[28]

And what about the Palestinians? The Hammer study cited above suggested close genetic ties between Jews and Palestinian Arabs, as well as Syrians. Lucotte and Smets also present data suggesting genetic similarities between Jews, Palestinian Arabs, and Lebanese—with all three groups exhibiting haplotypes VII and VIII. If Palestinian Arabs are descendants of Ishmael, son of Abraham, they might share Y-chromosome sequences, since they share a common patriarchal line. Abraham also fathered and passed his Y-chromosome on to Ishmael, as well as to Isaac. Isaac fathered Jacob, who transmitted that Y-chromosome on to his twelve sons, fathers of the twelve tribes. However, despite some similaries in Y-chromosome genetic sequences found by two studies, the Cohen Modal Haplotype (CMH), which appears to be a true Jewish genetic signature, is present only in very low frequency in Palestinian Arabs.[29]

Israelite Roots of Other Populations

As the most remote areas of the world become more accessible, people are being rediscovered who claim Israelite lineage, or practice customs that are distinctly Jewish in nature. Literally millions of people all over the world fit these categories. Could we and should we test them to see if they carry Jewish genetic markers?

A group of people residing in northeastern India, Burma, and Thailand claim to be descendants of the Israelite tribe of Menashe, one of the ten "lost tribes" expelled from Israel by the Assyrians in 722 B.C.E.

It is believed that this group eventually moved east into China where they remained for about 1,500 years. They then migrated once again, ending up in Thailand, Burma, and the Mizoram and Manipur provinces of India. These people, numbering in the millions, are known as the Shinlung, although thousands of them call themselves B'nai Menashe, or Sons of Menashe. Many practice a form of Christianity, however they maintain some practices that are reminiscent of Judaism. They have holidays corresponding to the major Jewish festivals, they practice animal sacrifice similar to that described in the Bible, and they wear garments resembling *tzitzit* (the ritual fringed shawl used in Jewish prayer). Their facial features are Mongolian, their practices are ancient Israelite, and their home is southeast Asia.[30]

Some of the secret Jews who hid their faith through the terrors of the Spanish Inquisition have resurfaced in Latin America, Spain, Portugal, New Mexico, and the Philipines, and are rediscovering long buried Jewish roots. "In Brazil alone there may be as many as 15 million people—more than the entire world Jewish population—who descend from Iberian Jewish exiles."[31] They have not practiced Judaism openly for more than five hundred years. As they learn of modern Jewish practices and their similarities to their own traditions, many of these crypto-Jews, who secretly maintained some Jewish practices, are beginning to adopt a more Jewish identity and even seek out their Jewish brethren in their own countries and in Israel. Some wish eventually to emigrate to Israel.

Odmar Braga, of Recife, Brazil, is one example of a secret Jew, born into a Marrano family. During the Spanish Inquisition, Jews who pretended to adopt Catholicism but actually adhered to Jewish practices in secret were dubbed Marranos (from the Spanish word for swine). Braga grew up in the 1950s, in a family that lit candles every Friday at sunset, refused to work on Saturdays, and fasted on Yom Kippur and Tisha B'av. His family refrained from eating pork or shellfish, they maintained special prayers and read from the Hebrew Bible.[32]

There is a conundrum associated with people of Jewish descent who have been isolated for so long: their practices differ from that of mainstream modern rabbinic Judaism—and their practices frequently involve incorporating beliefs and rituals of other religions. In addition, their lineages may be mixed with those of other peoples, putting into question their status as Jews. So, without conversion, most rabbis will not accept them as Jews.

Can genetic technology be of assistance in determining the veracity of the claims of Jewish lineage of these groups? Could DNA testing yield data that would resolve their status?

Rabbinic Views on Genetic Identity and Jewish Status

As indicated above, most rabbis do not accept as Jews groups with Jewish heritage who are only recently returning to the fold. The rabbinate in Israel has insisted on conversion of individuals from B'nei Menashe, and other groups as well, who desire to emigrate to Israel via the Law of Return. (The Law of Return grants Israeli citizenship to all legitimately recognized Jews from anywhere in the world.) Unless Jewish lineage is clear and incontrovertible, traditional rabbinic authorities have required a formal conversion process before accepting individuals or groups. Jewish DNA does not appear to be an issue one way or the other at this time. The rabbinic question of who is an authentic Jew has never relied on DNA data, and most likely this will not change among traditional Jews. Orthodox and Conservative Jews maintain that family lineage, through the maternal line, determines Jewish status; halakhic conversion is required for all others wishing to be considered Jewish. Reform Jews accept matrilineal or patrilineal descent or Reform conversion to establish Jewishness.

The Israeli government, by practical necessity, would need to be conservative in evaluating the Jewish status of newly discovered people of Israelite descent. Theoretically if all the B'nei Menashe wished to enter Israel, it could mean 2–4 million new citizens. And there are as many as 15 million other potential immigrants if descendants of Marranos in Brazil desired to return.

It is unlikely that any genetic test will, in the near future, be used as a Jewish barometer. After all, Judaism is a religion, not a race. Ethiopian Jews who were halakhically converted are no less Jewish than direct descendants of Aaron. However, as studies of Jewish genes are completed, more groups of people may be confirmed as descendants of Aaron and the twelve tribes, and the little State of Israel may be faced with absorbing millions of lost tribe members.

"Who is a Jew?" is clearly not a question that geneticists will be answering anytime soon. Rabbis will continue to be the ones to determine Jewish status. With regard to the Lemba, Rabbi Norman

Bernhard of the Southern African Rabbinical Association commented that "it may strengthen their historical claim that they come from Jewish origins, but it doesn't do anything to solve the halakhic complications." He further remarked, "I can invite them for dinner on Friday night, but I can't let them marry my daughter."[33]

Who Are You?

New techniques in DNA sequencing have provided us the wherewithal to determine, with high accuracy, paternity, maternity, and identity of human remains. Paternity is, of course, critically important in determining the status of a child within the Jewish community. Determination of maternity should not be an issue and would appear to be obvious, since most mothers are present for the births of their children. But, as we have discussed in chapter 2, egg and embryo donation can change that situation, since the birth mother may not necessarily be the genetic mother. Lastly, identification of human remains has been a subject of rabbinic writings since the times of the Mishnah, and it is still fraught with controversy. Accurate identification of dead bodies is of critical importance in Judaism (as in most other cultures). Correct identification is necessary to determine rites of mourning (such as who is to sit shiva—i.e., observe the seven days of mourning; who must say kaddish—the traditional prayer in honor of the deceased), inheritance of property, and to release a woman who has been widowed so that she may marry again.

With regard to paternity, rabbinic scholars have discussed the use of blood testing. The determination of A, B, and O blood groups, which has been available for decades, can be used to exclude someone from paternity but is not useful to definitively identify someone as the father. There are some interesting issues regarding halakha and blood testing. According to the Talmud, "Our Rabbis taught: There are three partners in man, the Holy One, blessed be He, his father and his mother. His father supplies the semen of the white substance out of which are formed the child's bones, sinews, nails, the brain in his head and the white in his eye; his mother supplies the semen of the red substance out of which is formed his skin, flesh, hair, blood and the black of his eye. . . ."[34] Based on this passage, which ascribes inheritance of blood from the mother alone, some rabbis have concluded that blood tests cannot be used to determine paternity.[35]

Clearly, this talmudic passage does not reflect our current understanding of how human heredity works. It is a reflection of ancient understanding of inheritance. Maimonides stated that we are not required to accept the scientific notions of the Sages, since their knowlege of science was not derived from Divine sources (such as the Written or Oral Law) but from prevailing (human) views of their times. Rabbi Shlomo Zalman Auerbach agreed, suggesting, "it is possible that the works of our Sages are not to be taken literally, and have nothing to do with blood types."[36] Nevertheless, there still are rabbis who insist that we cannot use scientific data that contadict views of the talmudic rabbis. Hence, those rabbis reject any application of scientific testing for paternity via the blood.

However, given that blood testing is accepted by many mainstream rabbis as a method of detemining genetic relationships, Professor Dov Frimer concluded the following regarding Jewish law and blood testing:[37]

1. If results of blood testing show paternity with a "high degree of probability, but not absolute verification," that would not be enough to obligate the man to pay child support, or to declare the child an heir.

2. The determination of the legitimacy of the child, for halakhic purposes, should not rely on blood testing. Even if the results "rule out" the husband as a possible father, it still does not prove from a halakhic standpoint that the woman has performed adultery and that the child is illegitimate. Rabbis go to great lengths to avoid classifying a child as illegitimate, since the status of *mamzer* is not just a social stigma in Judaism, it essentially excludes the child from many communal relationships—including marriage within the legitimate community. The status of *mamzer* also stigmatizes the descendants of that child. Rabbis have argued that these tests are never 100 percent certain, since mistakes can be made, so they tend toward lenient rulings in these cases in order to avoid labeling people as *mamzerim*.

3. A blood test could be sufficient to clear a person of a criminal offense. For instance, if a man has sexual relations with a woman, and it is suspected that she may be his daughter, the possibility of incest can be eliminated with a negative blood test.

Frimer's paper also discussed HLA testing, which, in 1989, was a recently developed procedure that proved to be much more useful and more accurate than A, B, O testing. HLA (Human Leukocyte Antigen) testing involves the identification of antigens, which are molecules that stud the surface of all cells. Every person has a unique set of these markers on every cell of his body. The markers are genetically determined; thus, HLA testing is a measure of genetic relationships.

HLA testing appears to be preferable from a Jewish standpoint for a number of reasons. First, since red blood cells do not have to be tested to determine HLA status, the problem some rabbis have regarding maternal versus paternal inheritance of blood becomes moot (so all rabbis should, theoretically, accept the test). Second, the accuracy of the test is much greater than 90 percent, and even approaches 100 percent with current technologies. Third, it can rule out relationships between individuals (as A, B, O testing can sometimes do) but also can positively establish relationships (which is beyond the capabilities of A, B, O testing).

HLA testing appears to have the support of many rabbis. The results of such testing could release a man from child support but should not be used to declare a child a *mamzer* (because there is always a remote chance that the test is incorrect). In fact, Justice Menachem Elon of the Israeli Supreme Court "advocated accepting H.L.A. test results as reliable evidence not only for excluding paternity, but also for positively establishing paternity." However, the HLA test should not be used "if there is a possibility that the results of such a test may affect the status of the child and cast a doubt upon his legitimacy."[38]

Since DNA testing may be analogous to HLA testing with regard to determination of genetic identity, rabbinic views on HLA testing could be extrapolated to help determine halakhic perspectives on the use of DNA technology. The accuracy of DNA testing approaches 100 percent, but, rabbis might correctly argue, no test is 100 percent certain. In DNA testing (as in HLA testing), human error and mechanical failure could affect the results. In the Nicole Simpson murder trial, the blood-stained gloves matched the DNA of defendant O. J. Simpson with a greater than 99.9999 percent certainty. And yet O. J. Simpson was acquitted. The jury based its decision, to a great extent, on the possibility of police tampering with the evidence (not lack of reliability of the test, but lack of reliability of the investigators). As with HLA testing, rabbinic reactions to DNA testing may also consider the possibility of human fallibility and thus disallow DNA tests. Rabbinic reactions to HLA testing would also suggest that rabbis might not permit DNA testing, or might disallow the results, if the data would lead to the identification of someone as a *mamzer.*

The use of DNA testing for identification of human remains has generated some controversial issues. Because of frequent terrorist activity directed against Israelis, authorities in Israel have learned all too well how to carefully collect scattered body parts, cast out of bombed buses and buildings. One of the grisly tasks of the Israeli Police is to identify remains and match up body pieces for interment. Grieving relatives need to know with certainty when their loved ones are the victims. The immense tragedy at the World Trade Center in 2001—where thousands of people were crushed beyond recognition—resulted in a situation where DNA testing was, in many instances, the only recourse for identification of remains.

Within traditional Judaism, it is particularly important to positively identify human remains and firmly establish the death of a married man, in order to determine that his wife has become a widow. That is because in Judaism, in order for a woman to remarry, there must be a religious divorce (a *get*) or definitive proof that her husband was deceased. Women whose husbands refuse to grant a *get,* or those whose husbands disappear, or die without recovery of a body, may be relegated to the status of *agunah.* An *agunah* is a "chained woman" who may never remarry in accordance with traditional Jewish law.

This issue of using DNA testing for identification of human remains is so new that few rabbis have made rulings on these issues. Rabbi Richard Address, Director of the Union of American Hebrew Congregations Department of Jewish Family Concerns explained that the Bioethics Committee of UAHC has not examined this issue yet. But based on Reform Judaism's stand, which does not insist on a halakhic basis for rendering decisions, Rabbi Address comments, "This agunah issue is not a prime concern for the majority of non-halachic Jews, who follow the dictum of *'dina d'malchuta dina':* the law of the land is the law." He continued, "This is not a 'do whatever you want according to the situation' ethic. It is, however, a belief that the context of a case helps decide how one chooses to apply fundamental Jewish values. . . ."[39]

Halakhic views on body identification can be traced back to the Mishnah: "They must not give evidence except from the face together with the nose, even though there be marks on its body or on its clothing."[40] This passage refers to evidence or testimony that is presented to permit a woman to remarry in the event of the death of her husband. The corpse must have an intact face and nose for there to be positive identification. The Mishnah goes on to present examples of testimony regarding a person's death. Some of these examples appear to be quite lenient, since the Mishnah attempts to ensure that a widow will be able to remarry. For instance, if children are overheard reporting a man's death, or if an echo, or an unseen voice is heard announcing the passing of a person, then the woman may be declared a widow and be permitted to marry.[41]

It would seem that DNA identification of a faceless corpse should be sufficient to release a woman to marry again. Modern techniques should be accepted for identification, as they have proven to be more accurate than traditional methods. "If the shift in emphasis from human testimony to scientific evidence is now to become a standard, this truly constitutes a revolution in halakhic thinking," notes Jay Levinson of the Division of Identification and Forensic Science, Israel Police National Headquaters. Levinson explains how "in victim identification the thrust moved from personal recognition to scientific findings, and from individual memory to often physical evidence."[42] In Israel, the rabbinic involvement in body identification has been reduced, as the role of the army and the police, using forensic evidence, has increased.

Ironically, it turns out that the religious courts are sometimes more lenient in their application of eyewitness accounts than the rigorous standards of forensic science. By considering DNA and other physical evidence as more reliable than eyewitness accounts, the police appear to be ruling on the stringent side, while the rabbis rule on the permissive side. Forensic science has determined that eyewitness accounts can be tainted by faulty memory or emotional reactions to trauma, thus police would much rather rely on modern techniques and forensic evidence.[43]

Rabbis from Ohr Somayach Institutions/Tanenbaum College in Jerusalem have ruled on the use of DNA technology for identification of human remains. A question posed to these rabbis on their website was, "I read that genetic testing was used to identify some of the people killed in the Tel Aviv bus bombing. Does genetic testing have Halachic validity?" Rabbis Moshe Lazarus, Reuven Subar, and Avrohom Lefkowitz accept the use of DNA testing to verify the death of a person for the purposes of proper burial. However, they draw the line at permitting the widow to remarry based on DNA evidence alone. They insist on using talmudic means (i.e., testimony) to determine the identity of the deceased, and they find genetic testing unacceptable. According to this responsum, "I asked Rav Chaim Pinchas Scheinberg, *shlita,* specifically about permitting a woman to remarry solely on the basis of genetic testing and about using those tests to resolve disputes that may arise because of inheritance. He said that there *is not* any ground in Halacha that would permit us to do so" (emphasis in original).[44]

Rabbi Moshe Klein of Bnei Barak, Israel, has also reportedly issued a halakhic decision regarding DNA evidence. Written together with Rabbis Wozner and Karelitz, the decision asserts that "DNA does not meet the strict halachic standards of certainty." Thus, DNA testing would not establish someone as a *mamzer* and should not be used to convict someone on criminal charges. In addition, these rabbis cautioned, "DNA cannot be used as sole evidence to determine death and allow a wife's remarriage."[45]

And what of the hundreds of Jews lost in the World Trade Center tragedy? Jay Levinson was rushed to New York shortly after the disaster; his expertise contributed to the efforts to identify human remains and—within halakhic bounds—to help declare Jewish victims as deceased. In this case, Levinson indicated, personal testimony would be critical for determination of whether a person was located at the site of the disaster. Fingerprints, dental records, distinct scars, and tattoos are considered top level signs, or *simanim,* to identify human remains; but most of the remains recovered would provide none of these signs. Personal property recovered on the site could also be used as part of the evidence to declare someone a casualty of the calamity. Even in this extreme situation, Levinson maintains, DNA testing should not be the only criterion.[46]

The reluctance to rely on DNA testing alone (albeit with an accuracy of more than 99.9999 percent) to conclusively identify human remains appears to fly in the face of previous Mishnaic interpretations. The Mishnah accepts a number of questionable and uncertain types of testimony in order to release widows so they may remarry. Now with the availability of DNA technologies it is possible, with a great degree of confidence, to declare someone dead even if the body is in

parts or highly decomposed. And some rabbis are rejecting the technology, or they recommend limiting its use. Ironically, if they accept
talmudic methods, they may be able to accept the testimony of an
"echo," or children, or even decomposed bodies which are unrecog-
nizable ("they did not recognize him, nevertheless they allowed his
wife to be wed again"[47]). It appears contrary to the halakhic princi-
pals that govern other decisions to reject identification by DNA—the
signature of human identity.

Implications of Jewish Genes

In a world where family and communal relations are static and un-
changing, and people stay in one place, genealogy is easy to study. In
the Jewish world, life has been anything but static. Communities have
split and merged, and people moved and settled and moved again.
The amazing thing is that Jewish identity has been maintained despite
the dynamic nature of Jewish communities throughout history.

How can gene studies illuminate the dynamic aspect of Jewish migration and composition of Jewish communities? These types of
studies can give us insight into history. They can confirm the written
and oral accounts of mass migrations, conversions, and dispersions.
In addition, gene studies can resolve some difficult questions with regard to family relationships and identification of human remains. All
of these issues have an impact on Jewish identity and on application
of halakha.

What other implications are there for genetic analysis using Jewish
genes? Could tests be developed to confirm Priestly descent? Kohanim
could be tested to determine if they truly deserve that status. Men
who test positive for "priestly genes" might be of Priestly descent,
however, the presence of these markers will not prove it. On the other
hand, men who do not carry the Y-chromosome markers can, theoretically, be ruled out as being direct patrilineal descendants of Aaron.
Should they be forced out of the Priestly fold? Some groups may decide to test men for Priestly markers to determine if they qualify for
the responsibilities and privileges, and if they must adhere to the constraints of the priesthood. Israelite and Levite men might also be
tested to determine if they are of Priestly stock, to determine if they
may marry a divorcee, or enter a cemetery, or even assume the rights

and privileges of the Priestly status. Biology may supersede the oral tradition and determine status in the Jewish community.

Leonard Nimoy, better known as Spock, originated the split finger greeting ("live long and prosper") on television's *Star Trek*. The actor has remarked that the idea for the gesture came from his childhood, while observing the Kohanim bless the congregation, using the Priestly split finger gesture, in an Orthodox synagogue in Boston. He recalls "the *kohanim* wailing their chant under their great *tallisim*, their hands extended toward the congregation, fingers splayed." Nimoy reminisced about his "fascination and . . . peeking in spite of my father's admonition." And he explained "how I introduced the salute into Star Trek and the Vulcan culture."[48] He clearly can perform the Priestly gesture (and has, hundreds of times, on TV). Perhaps Spock should be tested. Maybe he's a genetic Kohen, too.

CHAPTER TEN

Judging Genes

"ASHKENAZI JEWISH FAMILIES are needed to help scientists understand the biological basis for schizophrenia and bipolar disorder," read one advertisement, recruiting volunteers for biomedical research. These ads ran in newspapers throughout the country, prompting Rabbi Moshe Tendler to remark, "We need this like a hole in the head." Tendler explained that studies like these, focusing attention on only one population of people, will reawaken the idea that "Jews carry genes that are polluting the world. That's the basis of eugenics," Tendler explained. "If you have a [disease] gene, don't you owe it to society *not* to propagate that gene?"[1]

Once it is possible to screen for a variety of genes that determine different traits, ethical questions will abound. It will become critical to make judgments regarding those genes—whether they are genes that cause physical malformations, cancer, psychiatric conditions, or even genes for traits such as physique, stature, musical ability, or skin color. Genes may be judged as good or bad, and those who carry the genes likewise will be judged. Fetuses with lethal genes are already being selectively aborted. Some couples even selectively abort fetuses of the undesired sex. Is there value in the power to select for or against certain genes? Does this contribute in a positive way to society? What impact does it have on Jewish populations, in particular?

The advertisement quoted above was placed by the Department of Psychiatry at Johns Hopkins University, which is conducting a research study on schizophrenia and bipolar disorder. Their program in

Epidemiology-Genetics studies populations of people with psychiatric conditions. For more than a decade these studies focused on families of any ethnicity with more than one affected member, but in 1996 a major change was made in criteria for enrollment in the study: They began to focus mainly on "families of Jewish ancestry."[2] Dr. Cindy Hunter, a coordinator of the study, reported that they have enrolled at least 1,200 Jewish families in the research project.[3]

The idea that Jewish populations can serve as model groups for genetic research has irked many Jewish leaders. "Today, every time we read about 'bad genes' the term is hyphenated with 'Ashkenazi Jew,'" Tendler wrote. "We must not screen Ashkenazi Jews alone. Screen Scottish, screen Spanish. There should be a moratorium on publishing about Ashkenazi Jews until we include other populations in these studies as well. We must remember that the Holocaust was initially fueled by data supplied by medical professionals in Germany to purify the gene pool of the world—first of the mentally defective, then of the Gypsies, and finally of the Jews."[4]

However, the Johns Hopkins program explains their justification for using only Jewish populations. "Due to a long history of marriage within the faith, which extends back thousands of years, the Jewish community has emerged from a limited number of ancestors and has a similar genetic makeup. This allows researchers to more easily perform genetic studies and locate disease-causing genes." To illustrate the point, the Johns Hopkins website shows two bowls of candies. One bowl, which represents the general population, has brown, yellow, orange, blue, and green candies. A single red candy is mixed in with the others, but remains hard to discern in the potpourri of colors. The other bowl, representing the more homogeneous Jewish gene pool, has only yellow and orange candies. A single green one in the bowl is very easy to locate. Likewise, disease genes may be easier to identify in a more homogeneous genetic background.[5]

One major concern of critics of Jewish genetic testing is that Jews will be stigmatized as being less genetically fit than the general population. In a list of "Frequently Asked Questions" at the Johns Hopkins website, one such query is posed: "Are these disorders more prevalent in the Jewish population?" The answer provided appears to present contradictory statements: First, that "there is not sufficient evidence to indicate a higher risk for schizophrenia and bipolar disorder in the Jewish community." And second, that "there are some studies that suggest an increase in affective disorders, including bipolar disorder, in the Jewish community." However, the latter information, they admit, is not yet "adequately substantiated."[6]

Tendler also raised the concern that genetic screening could inhibit people from living their normal lives. He stated that some knowledge gained through testing might deter people from getting married or having children. "It would be a mitzvah to get tested only if there's therapeutic potential."[7]

Dr. Mandell Ganchrow, formerly the President of the New York–based Union of Orthodox Jewish Congregations of America, is also not convinced that Jewish genetic screening is a good idea. "Until we have mechanisms in place that will protect the information, we must be careful about advocating testing. We don't yet know how to secure information, how to use it, or how to protect the individual psychologically. . . . We find genetic disease in every subset of humanity. It is wrong to see Ashkenazim as inherently 'diseased.'"[8]

On the other hand, there are many who support the use of Jewish genetic studies. The value of such studies is that they may lead to progress in genetic diagnosis and therapy. Lois Waldman, Director of the Commission for Women's Equality, American Jewish Congress, in New York City, supports the contributions of Ashkenazi Jews in these studies. "My main concern is the blanket hostility toward testing," stated Waldman. "Those disposed to see a cloud in every silver lining have managed to cast doubt on research about the causes of and possible cures for genetic diseases prevalent among Ashkenazi Jews. There are good scientific reasons to focus on Ashkenazi Jews. The research is conducted by responsible scientists, many Jewish themselves, under stringent ethical guidelines. There are no credible reports of misuse of the data. . . . [I]t is a dishonest use of history to fail to recognize the differences between racist Nazi eugenics and responsible scientific research to relieve human suffering."[9]

The Johns Hopkins studies on schizophrenia and bipolar disorder are still in progress. The director of the study, Dr. Ann Pulver, is herself of Ashkenazi Jewish background. According to Dr. Cindy Hunter of Johns Hopkins, one of the project's coordinators, analysis of the data has not yet been completed, and the studies are yet to be published. So we do not have the answer yet as to genetic components of those psychiatric disorders in the Ashkenazi Jewish population.

Genetic studies of this sort could end up stigmatizing a population. With regard to Jewish groups, there is a foreboding associated with this kind of stigma. After all, the Nazis spent more than ten years trying to prove that Jews are genetically inferior. And now studies focusing on Jewish populations reveal Jewish Achilles' heels—numerous genetic variants that are associated with a variety of conditions and diseases.

Population Bottlenecks

In fact, disease genes are present in all populations. It is only because the Asheknazi Jewish population has been cloistered for hundreds of years and because the group has been subject to such scrutiny that it seems as if there are more disease genes in that group. In addition, Ashkenazi Jews have gone through what geneticists call a "population bottleneck." According to geneticist Ricki Lewis, "A population bottleneck occurs when many members of a group die, and only a few are left, by chance, to replenish the numbers." It can also occur when a small group of people are isolated from other groups, for instance on an island. The Pingelapese people, who inhabit islands in Micronesia, exhibit 4–10 percent incidence of Pingelapese blindness. Genetic studies have suggested that this group of people arose from a population bottleneck. Following a devastating typhoon in 1780, only nineteen Pingelapese people were left, who served as the founders of the present-day Pingelapese. Their small original number, their isolation, and subsequent inbreeding, have led to a high frequency of one form of blindness in the modern-day Pingelapese.[10]

In the case of Askhenazi Jews, a population bottleneck may have occurred as the result of frequent and repeated pogroms and massacres some 350 years ago. Lewis writes that "Human wrought disasters that kill many people can also cause population bottlenecks—perhaps even more severely, because aggression is typically directed at particular groups, while a typhoon indiscriminately kills whoever is in its path." Lewis recounted the Chmielnicki massacres that began in 1648. Within a few years, "thousands perished with only a few thousand Jewish people remaining." Their numbers grew from this founding population to several million by 1939. "But the Chmielnicki massacres, like others, changed allelic frequencies and contributed to the high incidence of certain inherited diseases among people of eastern European Jewish heritage."[11]

According to Dr. Kenneth Offit, Department of Human Genetics, Memorial Sloan-Kettering Cancer Center, the rise and fall of the world Jewish population resulted in "founder genes" appearing in higher proportions in Jewish groups than in the general population. As a result of assimilation, crusades, and pogroms, a peak population of 8–12 million Jews in the year zero of the Common Era plummeted to fewer than 500,000 in the 1700s. Then, that small "founder population," where particular genes may have been found, underwent explosive expansion from 1720 to 1930. Since most people in that population married within the group, their millions of Jewish descendants living in Europe in the twentieth century ended up with a high proportion of them carrying specific forms of genes from the founder group. Some of those genes may have been disease genes such as Tay-Sachs and the BRCA1 and BRCA2 mutations associated with breast cancer.[12]

Other homogeneous populations have also been identified as carrying higher frequencies of certain forms of genes. For instance, the sickle cell anemia gene occurs in 1 out of 400 U.S. blacks but is very rare in white people. Congenital hypothyroidism occurs at about a 1 in 5,000 frequency in whites, but is rare in black populations. Congenital adrenal hyperplasia is found in 1 in 680 Yupik Eskimos, but occurs in only 1 out of 12,000 whites. Other populations that have been used for genetic studies because they are fairly homogeneous by virtue of long-term genetic isolation include Icelandic people (high incidence of arthritis), rural residents of a plateau in Costa Rica surrounded by high mountains (high incidence of manic-depressive illness), and citizens of the small Dutch fishing village of Bunschoten (high incidence of a rare liver disease). Each of these populations grew from a small founder population to tens or hundreds of thousands within a few hundred years.[13]

Non-Ashkenazi Jewish Genes

Non-Ashkenazi Jews also carry genes that are unique and more prevalent in their populations. The Sephardim arose from populations of Jews who lived under Moslem rule in Spain, Portugal, and the North

African coast. After Sephardic Jews were expelled from Spain in the fifteenth century, a number of distinct Sephardic subgroups developed in communities, many of which were located along the northern and southern Mediterranean coast, and in the Western Hemisphere. Their dominant language continued to be Ladino.

Over the two-thousand-year diaspora, Oriental Jews dispersed and established communities in the Middle East and Asia, including places such as Iran, India, Kurdistan, and Afghanistan. The dominant languages in these communities were Arabic, Persian, and Judeo-Arabic.

"Many subgroups have evolved from Oriental Jewry with distinct environmental and genetic features," explains Dr. Richard Goodman. Since Sephardic Jews also dispersed into isolated communities, more-or-less distinct subgroups emerged from their ranks.[14]

Because of the geographic history of non-Ashkenazi Jews and the relative isolation of specific subgroups, populations from different areas now exhibit characteristic genetic disorders. For instance, among Oriental Jewry, Jews from Iran and Iraq have a higher incidence of pituitary dwarfism, type II; Jews from India exhibit higher frequency of ichthyosis vulgaris; and Jews from Yemen have an increased rate of celiac disease and PKU. Among the Sephardim, Libyan Jews have an elevated rate of cystinuria; for Moroccan Jews, ataxia telangiectasia and familial deafness are more prevalent; and Tunisian Jews exhibit higher rates of selective vitamin B_{12} malabsorption. One disease that all Sephardic Jews share at higher incidence than the general population is familial Mediterranean fever, or FMF.[15]

So if non-Ashkenazi Jews also have genetic flags, why are most research projects directed toward Ashkenazi populations? First of all, Ashkanazim make up 90 percent of the Jewish population in the United States, where much of this research is carried out; so it is simply easier to get more subjects from the Ashkenazi Jewish group. Goodman reported that in 1986 Ashkenazim accounted for 82 percent, of world Jewry, the Sephardim represented 11 percent, and the Oriental Jews made up 7 percent.[16]

In Israel, the Ashkenazi and non-Ashkenazi populations are more evenly distributed, and they also have intermarried; thus in Israeli studies it is more difficult to focus only on Ashkenazi groups. In 1986, Goodman reported that only 47 percent of Israeli Jews were Ashkenazi.[17] More recent data are difficult to analyze based on group origin, since now a significant proportion of Israeli Jews are Israeli born, and a large percentage of them also have Israeli-born parents. Data from the Israeli Government Central Bureau of Statistics show that in 1948 only 37.8 percent of Israel's population was born in Israel; that has increased to 63.2 percent in 1998. The category "Israeli born and father born in Israel" rose from 5.5 percent in 1948 to 27.7 percent in 1998. Of the remaining Jews in Israel (72.3 percent), 1998 figures show that about 21 percent are from countries with Sephardic populations, 14 percent from countries with predominantly Oriental Jews, 37 percent from countries with predominantly Ashkenzi populations, and 1.3 percent from Ethiopia. (The large influx of Russian Jews (576,000 from 1990 to 1995) into Israel has added significantly to the Ashkenazi population.)[18]

The most important reasons that Ashkenazi Jews have been the subject of greater genetic focus and scrutiny are the homogeneity of the group compared to others and easy access to a large population willing to participate in medical research studies. "When you raise the flag of science, a Jew salutes," quipped Rabbi Moshe Tendler.[19] However, the Jewish population in Israel is not as homogeneous as Jewish populations in the United States; thus, the large populations of non-Ashkenazi Israeli Jews also are considered in many Israeli genetic studies. It is important to note that since Jews of Sephardic and Oriental origin have numerous distinctive subgroups in their populations, the greater heterogeneity of non-Ashkenazim makes genetic studies on them much more complex.

Recessive vs. Dominant Genes

Genes that occur in higher frequency in Ashkenazi Jewish populations include those that are inherited as recessive genes and cause eight serious and frequently lethal diseases (Tay-Sachs, Canavan, Neimann-Pick, Gaucher, cystic fibrosis, Fanconi anemia, Bloom syndrome, and familial dysautonomia—see chapter 6). For all of those diseases, carrier status can be determined in couples before conception, and if the couple so desires, pregnancies can be tested and terminated. Tay-Sachs, Canavan, and Niemann-Pick are lethal within the first few years of life. The other five diseases are more variable in severity. For diseases determined by recessive genes, one copy must be inherited from each parent in order for the offspring to express the disease.

But some genes work differently. Numerous conditions and diseases result from dominant genes. It only takes one copy of a dominant gene to determine a dominant trait. Thus if one parent has the gene, each child has a 50 percent chance of inheriting the gene. The genes may vary widely in expression, ranging from no indication of traits to full expression. The properties of such genes that determine expression are called penetrance and expressivity. Penetrance refers to the frequency with which a gene that should be expressed does, in fact, express itself. Expressivity refers to the *degree* of expression.

One example of a dominant gene that varies widely in expression is the gene for dimpled cheeks. The trait of cheek dimple is, of course, not a disease—it is an attractive facial feature. If a person with the trait carries one copy of the dominant dimple gene, her child would have a fifty percent chance of inheriting that gene. Children who inherit the gene can display a wide range of facial traits or phenotypes. The penetrance of that gene is less than 100 percent, thus some children can inherit the gene but not have dimples. (They can still transmit the gene to the next generation, which may or may not have dimples.) The expressivity is varied and can manifest itself as one or two dimpled cheeks and as deep or less pronounced dimples. For many genes where penetrance or expressivity is less than 100 percent, the factors determining the degree of expression may include environmental influences, diet, modifier genes, or other factors yet to be identified.

Some conditions or diseases determined by dominant genes are not expressed until later in life. These types of traits may or may not be lethal. Their range of expression may vary from no expression to the most severe symptoms, which in the case of lethal genes means death. Huntington's chorea is an example of a disease determined by inheritance of one copy of a dominant gene. It is a progressive neurodegenerative disease that appears later in life.

Genetic Dilemmas Relating to the BRCA Genes

Hundreds of years ago in an Ashkenazi Jewish community, a DNA mutation occurred that would affect hundreds of thousands of people in the twentieth and twenty-first centuries. A Jewish baby conceived from a mutated sperm or egg was the ancestor who transmitted a faulty version of the breast cancer gene BRCA1 to his or her offspring and to dozens of subsequent generations. That genetic mutation was recently discovered and dubbed "185delAG."[20] Female descendants who inherit 185delAG have a significantly elevated chance of contracting breast cancer.[21] Another unrelated event, in another sperm or egg, at another time and place in the Ashkenazi community, led to the mutation of a second form of breast cancer gene, BRCA2. Many descendants of that common ancestor have inherited the mutant gene, called 6174delT. Those individuals also are at greatly increased risk for breast cancer.[22] Today the incidence of those BRCA1 and BRCA 2 mutations is estimated to be 1 in 50 Jews of Eastern European ancestry.[23] In the United States that translates to about 100,000 Jews, or 50,000 women who may carry one of those mutant genes. And those two specific mutations do not occur at any significant level in the non-Jewish population. (Although other types of mutation in BRCA1 and BRCA2 do exist in the general population, they are present at a much lower frequency.) According to the National Cancer Institute, approximately 1 percent of Ashkenazi Jews carry the unique mutation 185delAG, whereas it is estimated that only 0.1 percent of the general U.S. population have *any form* of BRCA1 or BRCA2 mutation.[24]

The mutations, which occurred so many generations ago, changed the genetic alphabet of two essential genes—genes that are supposed to suppress or turn off cell growth. BRCA1, a gene located on chromosome 17, and BRCA2, located on chromosome 13, are normal genes required to prevent cells from growing out of control and becoming cancerous. The mutations involved the loss or deletion of one or two of the genetic "letters." All DNA is composed of four chemical building blocks, or bases, which are abbreviated A, C, G, and T. Those four chemical "letters" are arrayed in a specific order to spell out the complete genetic code of every creature. The complete genetic code for a human contains about 3 billion letters, arrayed on 23 pairs of chromosomes. In the mutation "185delAG" the letters A and G are missing in the 185th position of the BRCA1 genetic code. The "6174delT" mutation involves the deletion of T in position 6174 of the BRCA2 gene.

BRCA1 and BRCA2 can be inherited and transmitted by men or women in the same manner. However, the genes are not expressed in men and women in the same way. Although men who inherit a mutant BRCA1 gene exhibit a moderate increase in risk for prostate or colon cancer, most men carrying the gene remain cancer-free.[25] Men who inherit a mutant BRCA2 gene *are* at higher risk for breast cancer. However, since male breast cancer occurs at 1 percent the rate of female breast cancer, the disease is still rare even in men who have the mutant BRCA2 gene. Most startling is that women who inherit mutant forms of BRCA1 may have up to a 46 percent risk of contracting breast cancer and a 44 percent risk of contracting ovarian cancer in their lifetimes.[26] (This compares with an overall 12 percent risk of breast cancer and 1 percent risk of ovarian cancer in the general female population.)

The discovery of specific mutations in breast cancer genes found predominantly in individuals of Eastern European Jewish ancestry has led to an unprecedented and devastating dilemma. Jewish families where one or more women have had breast cancer may need to consider the possibility of a genetic mechanism of inheritance for that disease. With that possibility comes the question of whether or not to genetically screen unaffected members, and what to do about those individuals identified as carriers of the cancer gene. Some women presented with a very high chance of contracting breast cancer opt for the most extreme reaction: prophylactic mastectomy, or complete removal of the breast tissue. And even young women in their twenties, with no symptoms of cancer whatsoever, have elected this procedure. Others would rather not even know if they have the gene and refuse to be screened. There are a wide range of reactions and approaches to the situation.

It is important to understand that inherited breast cancer genes probably cause no more than 10 percent of all breast cancers.[27] (Other cases of breast cancer may be attributed to environmental factors, lifestyle, hormonal factors, or other inherited traits. That explains why women who do not have the mutant BRCA1 or BRCA2 genes still have about a 1 in 9 chance of contracting breast cancer by age 80.) Within the 10 percent of all inherited breast cancers, the presence of one of those BRCA mutations does not automatically guarantee that one will contract the disease. In fact, the two genes in question, BRCA1 and BRCA2, act like recessive genes. That means that there must be 2 copies of the faulty BRCA gene to cause cancer. This fact should make cancer a very rare event. (If 2 out of 100 people carry it, then 2/100 carriers who marry 2/100 carriers = 4/10,000 or 1/2,500 born who have two copies.)

The only problem is that inheriting two mutant copies is not the only way to get two mutant copies. A person can inherit one mutant copy of the gene, and the second, normal copy might undergo a chemical change and mutate, resulting in two flawed copies; and this is a recipe for cancer. Is the BRCA gene likely to mutate? Most genes are susceptible to mutation. Since we have two copies of each gene, any given mutation usually does not have a great impact on any individual adult cell; after all, we have a backup copy of most of our genes. Even if an occasional adult cell dies because of having two faulty copies of an important gene, it is likely to go unnoticed in a multicellular organism (we have about a trillion cells in our adult bodies). However, the breast cancer gene, in its normal form, is such a critical gene because it controls cell growth. When that cell growth gene is not functioning properly, the cells begin to grow out of control, and cancer is the result. Even if mutation is extremely rare, occurring in less than 1 in a million cells, since there are millions of breast cells in which the mutation can occur, over the lifetime of a woman that mutation in the backup gene *is* likely to occur. And mutations do occur with greater frequencies in certain environments. Smoking, dietary factors, environmental toxins, and exposure to radiation all increase the frequency of mutation. If an individual already has one faulty copy of the gene, a mutation in the backup copy will have a dramatic and possibly fatal impact on that person.

The New York Breast Cancer Study is "a multi–institutional research project studying the role that specific genes, called BRCA1 and BRCA2, play in the development of breast cancer in women of Jewish ancestry."[28] The study, which is being conducted at ten medical centers in the New York metropolitan area, is directed by Dr. Mary-Claire King and Joan Marks. King, a researcher at the University of Washington in Seattle, led the research team that first described the BRCA genes.

The goal of the study, which began in 1996, is to collect data from 1,000 Jewish women with breast cancer. Three founder mutations are being studied: the two genetic deletions described above—185delAG in BRCA1, and 6174delT in BRCA2; and an insertion mutation, 5382insC, where one extra genetic code letter, C, has been inserted in the BRCA1 gene.

Jessica Mandell, a Certified Genetic Counselor and the Research Coordinator of the study, explained that this study is focusing on environmental factors as well as genetics. Some environmental factors being monitored include: family origins; DES exposure; lactation; use of tamoxifen or raloxifene; social, educational, and occupational history; urban, suburban, or rural residence; and diet. The investigators hope to further understand the BRCA mutations, "to determine lifetime and specific risks, to determine if inheritance through females differs from inheritance through males, and to identify environmental or lifestyle factors associated with breast cancer," reported Mandell.[29]

The genetic dilemmas involving breast cancer are different from those encountered in families with the Tay-Sachs, Canavan, or other lethal genes, for several reasons. First of all, in the general population, a woman who lives to the age of eighty has a one in nine chance of contracting breast cancer. So this disease affects a very large proportion of the population (especially if you consider the impact on each victim *and* her family). Second, the decisions to be made are not clear cut. Inheriting either of the mutant BRCA genes can increase the likelihood of contracting breast cancer to more than 40 percent. (And, in some families with particular variants of the gene, the risk can be as high as 85 percent.) However, although the risk is high, it is still not 100 percent. Some women have elected prophylactic mastectomy and ovarian surgery, preferring to lose their breasts and ovaries rather than risk cancer. That is not a perfect solution, since modern surgery still cannot remove every last cancer cell, and some cases of breast and ovarian cancer have been reported in women who went through those procedures. Others elect vigilant medical monitoring, including regular mammograms and frequent breast self examination, and they take their chances on contracting cancer.

Results of genetic testing can have great value if presented along with appropriate genetic counseling. Negative test results have helped women avoid unnecessary surgery. Positive test results have helped women make informed decisions on what steps they can take to increase their chances of a long and healthy life.

But most medical practitioners do not recommend universal testing of all Ashkenazi Jewish women for the BRCA genes. Doctors are concerned that negative tests will result in complacency when there should still be vigilance. After all, most cases of breast cancer are not due to inheritance of cancer genes. When there is clear evidence of a genetic link to breast cancer within the family, testing for mutant BRCA1 and BRCA2 genes may be recommended. However, some people are concerned that positive test results can lead to genetic discrimination by insurance companies or employers. There is concern that women who test positive for BRCA1 or BRCA2 mutations may become uninsurable because of a previously existing medical condition.

Fortunately, legislation to protect genetic privacy has been enacted in many states. At the national level, the Health Insurance Portability and Accountability Act (HIPAA, also known as the Kennedy/Kassebaum law, after its cosponsors in the Senate) makes it possible for people to maintain coverage when they change insurance or jobs, and assures that many uninsured people who start a new job are eligible for employer-provided group coverage, even if there is a preexisting medical problem.[30]

Rabbinic Views on Genetic Dilemmas

Rabbinical sources are conflicted about the choices facing women and families where BRCA mutations occur. Rabbinic consensus supports research that will lead to therapeutic treatments or cures of disease. As Rabbi Elliot Dorff writes, "The Jewish tradition would certainly not object to such research; it should actually push us to do as much as we can to learn about these lineages so that hopefully one day soon we can help people avoid cancer or, failing that, cure it. This attitude follows from the fundamental Jewish approach to medicine, namely, that human medical research and practice are not violations of God's prerogatives but, on the contrary, constitute some of the ways in which we fulfill our obligations to be God's partners in the ongoing act of creation."[31]

Dorff states that when gene therapy becomes available to treat or repair cancer genes Jews will be obliged to use these treatments, to the extent of repairing mutated genes even before any sign of cancer appears. "Jews have the duty to try to prevent illness if at all possible and to cure it when they can, and that duty applies to diseases caused by genes just as much as it does to disease engendered by bacteria, viruses, or some other environmental factor."[32]

The idea of judging genes has already been raised in chapters 6 and 7, with regard to fetal screening. Some rabbis permit abortion of fetuses when lethal diseases are present—especially those diseases, such as Tay-Sachs and Canavan disease, where there is 100 percent lethality within a few years of birth, and mental anguish of the family is severe. The question of how to address choices regarding mutations leading to cancer would be more controversial. How far may we go in the process of judging these genes? For instance, would it be permissible for a couple to abort a baby who carries the BRCA mutation and thus would have a high risk of cancer?

"The one clear conclusion is that, unlike the case of Tay Sachs, genetic testing should *not* be used to justify aborting a fetus carrying the BRCA1 mutation," states Dorff. "After all, the incidence of cancer among those who inherit the mutation is not 100 percent, and the age at which the disease begins to manifest itself in those who do contract the disease varies widely, with onset usually not until the thirties or forties. By that time the woman will have lived a considerable portion of a normal life span and may even have children of her own." In addition, argues Dorff, decades into the future when this newborn is at risk for cancer, it is likely that more effective treatments and cures will be available.[33]

Religious concerns also include the question of obligation to divulge the genetic information. "Does that person's fiancé(e) have a right to know about the gene in his or her intended mate?" asks Dorff.[34] "I think they do, but the answer to this is not as obvious as one might suppose. The Jewish tradition places a premium on truth and honesty in speech, business practices, and in personal relations. . . . [However,] Jewish sources recognize that there are some times in life when tact should take precedence over truth, specifically, when telling the truth will accomplish no important end and will make someone feel bad, diminish that person's hope, or cause strife. Truth is thus a critical value, but not an absolute one."[35]

On the question of disclosure to potential mates, Ari Mosenkis cites Rabbi Judah the Pious, author of *Sefer Chasidim*, who held, as Mosenkis explains, that prospective mates "are obligated to reveal physical defects of illnesses, and failure to do so results in a voidable marriage." According to the esteemed talmudic scholar Rabbi Israel Meir ha-Kohen (Kagan), the author of *Chafetz Chaim*, a third party, such as a parent or matchmaker, may also disclose information regarding medical condition, and this does not violate the biblical prohibition against gossiping (Lev. 19:16). He emphasizes that a third party is permitted, but certainly not obligated, to reveal such information.[36]

The medical point of view regarding disclosure is that family and friends should be told when a cancer mutation is detected, according to Elsa Reich, Certified Genetic Counselor and Clinical Associate Professor of Pediatrics, Human Genetics Program, New York University School of Medicine. "It's not the 'Big C' it used to be," Reich said, referring to old taboos regarding cancer. Since the BRCA genes are passed through men also, it is important to consider the male relatives who may have inherited the gene and can pass it on to their children. Relatives should know about the presence of this mutation in the family so that they can be tested, if they wish, and be scrupulous about medical surveillance.[37]

When a suitor learns his fiancée carries a BRCA mutation, is he permitted to marry her? According to Rabbi Immanuel Jakobovits, "Most important from the eugenic point of view, is the ruling that one should not marry into a leprous or epileptic family. . . . According to Rashi, any (hereditary) disease is evidently included in this category."[38] Jakobovits explained that in early talmudic times prospective brides would visit the public baths so that the groom's female relatives could determine if she had any physical abnormalities. "Under certain conditions, the discovery of physical defects or ailments in either the wife or the husband even entitles the other party to dissolve the marriage."[39]

Rabbi J. David Bleich describes the talmudic recommendation to avoid marrying into a family of lepers or epileptics as "perhaps the oldest recorded item of genetic counseling."[40] This type of stigmatized family is defined as one in which three members have exhibited the disease. However, emphasizes Bleich, if a marriage into such a family does occur, once such a marriage is in place, there still remains an obligation to produce offspring, even though there is a serious risk of deformity.[41] The cardinal mitzvah to "be fruitful and multiply" (Gen. 1: 28) applies even to couples who are likely to transmit serious diseases to their children.

Other issues addressed with regard to correcting genetic defects include the consideration of what types of gene therapy may be permitted in the future, when genetic cures become available. For instance, a faulty gene could be corrected in each afflicted adult, or in the fetus— long before it develops cancer, or in the germ cells, making the family genetic heritage free of that mutation for future generations. Dorff asks if it is proper to tamper with the genetic destiny by fixing the germ line: "If we were to change not only this mutation but the germ line of all present inheritors of BRCA1, have we stepped over the bounds of our legitimate powers to cure and changed ourselves into virtual gods, or are we simply and legitimately preventing disease more effectively?" He responds that eradicating a sickness is clearly permitted. And eradicating the root of the disease would likewise be acceptable. "I do not see any reason why it should be permissible to cure a given disease in one particular fetus and not in that fetus's future offspring as well. On the contrary, since sickness is degrading, it would be our *duty* to cure the disease at its root if we could, so that future generations will not be affected."[42]

Mosenkis poses the question of whether mass screenings are permitted. "Mass screening for BRCA may actually be halachically obligated as not screening may constitute a potential danger to life, or *sakanat nefashot.*" On the one hand, argues Mosenkis, people who test negative will reduce their mental anguish. Those who test positive may benefit as well since uncertainty can be particularly stressful for some individuals—even more stressful than a clear positive result. And a positive test will encourage vigilant medical surveillance. On the other hand, admits Mosenkis, a woman who tests positive may experience fear, despair, depression, and concern about herself and her family being stigmatized.[43]

In comparing BRCA screening and Tay-Sachs screening, Mosenkis points out that Tay-Sachs is invariably incurable, but at least screening programs can be used to carry out selective matchmaking to avoid pairing two people who carry that disease. This approach is used very effectively in ultra-Orthodox communities where most marriages are arranged by a third party. He concludes that the benefits outweigh the risks in screening for Tay-Sachs. However, with regard to BRCA, screening can only lead to prophylactic bilateral mastectomy, or hormone therapy, or increased medical surveillance. None of these approaches is 100 percent effective in preventing the disease. Thus the benefits may not outweigh the risks.[44]

The risks of knowledge refer, in this case, to mental anguish. In a Midrash based on Isaiah, Hezakiah said to Isaiah, "Even if you see (him) [a sick person] about to die, Do *not* say 'set your household in order' lest [the patient's] mind faint." In other words, it is not always prudent to provide a patient all information, because they may give up all hope. "It is sometimes better to lie to the terminally ill person than to reveal his true condition if the truth would produce mental anguish," concludes Mosenkis. Some rabbinic sources are concerned that, for a person who is ill, bad news will cause despair and depression and his condition will worsen. However, women who test positive for the BRCA mutation are frequently still in good health. Hence mental anguish in that case should not have serious physical implications. However, Mosenkis cites Rabbi Moshe Tendler's opinion that "stigmatization and discrimination" are sufficient reasons to discourage mass screenings.[45]

Other Cancers

The BRCA2 mutations also appear to elevate risks for other cancers, such as prostate cancer, pancreatic cancer, gallbladder and bile duct cancer, stomach cancer, and malignant melanoma. These types of cancers are elevated in men and women who carry the BRCA2 mutation. In women who had already developed breast cancer there is a more than 50 percent risk of a second breast cancer in the remaining breast, and an almost 16 percent risk of ovarian cancer by age seventy.[46]

In addition, other genes found in high frequency in Ashkenazi Jews have been shown to be associated with increased risk of cancer. A mutation in the adenomatous polyposis coli gene, APC 11307K, was detected in 28 percent of Ashkenazi Jews with a family history of colorectal cancer, and 6 percent of all Ashkenazi Jews. A control group of two hundred non-Jewish people did not exhibit this defective gene.[47]

Perfect Pitch: Jewish Musical Genes?

Ashkenazi Jews, as a handy group for genetic testing, will continue to learn of their unique genetic legacies. As in many groups that have been genetically isolated for some time, some of those genetic legacies include genetic defects and diseases. However, genetic legacies can also determine unique positive traits and qualities valued in society. For instance, a recent full-page advertisement, run in national magazines by "America's Pharmaceutical Companies," shows a young girl sitting in front of a piano. The caption declares: "Music runs in her family. So does cancer. Like a recurring theme in a Bach concerto, hereditary disease strikes generations of families. . . ." In terms of Jewish genetic heritage, perhaps cancer and other genetic diseases do "run in the family." But the news is not all bad. Positive genetic traits, such as musical ability, also run in Jewish genetic lineages.

A very provocative study is underway to determine the genetic basis for one aspect of musical ability—perfect pitch—and Ashkenazi Jews are important subjects in this study. Perfect pitch is a trait not associated with health problems; nor is lack of perfect pitch a reason to be stigmatized. It is of interest biologically, since understanding the genetic basis for perfect pitch may provide insight into how the brain processes sounds.[48]

Perfect or absolute pitch is the ability to recognize and identify musical notes without any source of reference. For instance, a person with perfect pitch would be able to name the notes for a touch-tone phone, a bird call, or a ringing bell; whereas someone with relative pitch would need to hear a reference note, such as middle C, played in order to identify another note. Siamek Baharloo, Jane Gitschier, Nelson Freimer, and colleagues hypothesized that perfect pitch occurs as a result of both genetic and environmental contributions. A person may have the genetic predisposition for perfect pitch but never develop the ability unless he or she is exposed at an early age to musical training. Ashkenazi Jews, it appears, have a higher incidence of perfect pitch than the general population.[49] Is this due to genetics, or environment, or both?

Joanna Raboy, a vocal music teacher from East Brunswick, N.J., reported on her participation in the study. "In 1997, I participated in a San Francisco State University genetic study on Ashkenazi Jews who have perfect pitch. Researchers were looking for a genetic link among people with perfect pitch, a phenomenon that occurs in less than one half of one percent of the general population. A researcher from the university tested my pitch and drew my blood to compare to samples drawn from others in the study."[50]

The Shaham family also participated in the study. The family includes virtuoso violinist Gil Shaham, his sister Orli Shaham, a concert pianist, and his brother Shai Shaham. All three siblings have perfect pitch. Among the general population, 1 in 2,000 has perfect pitch, but the rate among musicians is 15 percent. And among classical musicians, there are a number of Ashkenazim said to have perfect pitch, including the Shahams, Vladimir Horowitz, James Levine, and Michael Tilson Thomas.[51]

Of 612 musicians studied, 40 percent of those who began musical training before age four developed perfect pitch. However, among those whose training started after age nine, only 3 percent developed perfect pitch. This would point to an important role for the environment in determining perfect pitch. Alternatively, perhaps those with greater inborn musical ability tend to have their musical training start at an earlier age. In support of a strong genetic contribution to perfect pitch, although only 14 percent of those without perfect pitch reported having a first-degree relative (mother, father, sister, brother, child) with that skill, a startling 48 percent of those with perfect pitch reported that they had a first-degree family member with that skill.[52]

The studies that are ongoing are now focusing on families, like the Shahams, where more than one member has developed perfect pitch. Referring to those with absolute pitch as "AP-1," the most recent study reports that when the siblings of AP-1 subjects are tested, data indicate that "AP-1 aggregates in families." Just as families with several carriers of BRCA mutations have been used to map the genes and elucidate the role of BRCA in development of breast cancer, AP-1 subjects and their families may enable investigators to map the genes for absolute pitch.[53]

The investigators hope that by studying the trait in Ashkenazi Jews they can identify a single gene responsible for the ability, since Freimer believes that "whatever genetic predisposition there is for perfect pitch, it is likely that it will be identical in a majority of Ashkenazi Jews."[54]

Future Implications of Judging Genes

The specter of genetic diseases in the Jewish population has reportedly led one Jewish woman, in her quest to become a single mother, to specify the use of gentile sperm for her artificial insemination procedure. Her reasoning was that she wanted to avoid the cancer genes. Ironically, she chose the right type of donor with regard to rabbinic recommendations. In such a situation a gentile sperm donor is preferred by traditional rabbinic *poskim* (adjudicators), since there is no possibility for such a father to produce a *mamzer* (illegitimate child). It is also an appropriate choice from a religious perspective, since it is less likely that the child of this union will marry his or her half-sibling. From a genetic point of view it may make sense to avoid a Jewish sperm donor in order to avoid matching up two carriers of the same recessive disease (e.g., Tay-Sachs, Canavan, Familial dysautonomia). But any donor, of any ethnic background, will bring along his own genetic baggage; in the case of a dominant gene, there would be a 50 percent chance of that gene being transmitted to and expressed in the child. In addition, no matter what the genetic background of the father is, if the mother carries a BRCA mutation there still is a 50 percent chance her offspring will inherit it. And even if a child inherits only one mutant BRCA gene, the second copy can readily mutate, bringing cancer in its wake.

In some respects, genetic studies may provide more fodder for people who look for any excuse to stigmatize the Jewish population. However, if these studies are done with the right motives, and their results interpreted properly, there should be no basis for Jews to be stigmatized by such knowledge. It must be emphasized that many other groups have their own unique genetic flaws, foibles, *and* talents. If done properly, studies on all unique groups will contribute invaluable data to help us understand genetic mechanisms of disease and of human development.

There is no easy solution to the devastating dilemma of Jewish genetic flaws. We must learn more about the genes involved and the inheritance and action of those genes. We also need to understand how mutations occur, and how the environment can influence the rate of mutations. Many states have passed legislation to protect patients from genetic discrimination. That type of legislation must be universal, and should be continually reworked and strengthened to guarantee that we cannot be classified as uninsurable because of genetic defects. We must not be forced to be secretive about or ashamed of our genes. After all, no one is genetically perfect.

And there is hope that eventually we will not only understand breast cancer, mental illness, and hundreds of other defects, but we will be able to predict who will get those diseases, and cure them. The true cure for breast cancer will likely not be surgery, chemotherapy, or radiation, rather it will likely be based on the correction of the very genes that caused all the problems in the first place.

Ironically, genetic bonds, whether of a positive or negative nature could forge and help maintain other bonds within a population. The more we learn about the common genes in the population, the more we can appreciate other commonalities. To be sure, the "Chosen People" are not all genetic descendants of the original band of Israelites, since people who convert into the faith do not share that common genetic background. But for many in the Jewish population it appears that Jewish flesh has been marked by more than the *mohel*'s knife.[55] The indelible DNA imprint has left its lasting mark.

CHAPTER ELEVEN

Kosher Pork: Brave New Animals

WHEN A LOCAL kosher fast-food restaurant opened a while back, they offered a kosher "cheeseburger" (made with ersatz cheese), and they dubbed it a "*gevina*-burger." Since I grew up observing the kosher dietary rules, which, among other restrictions, means never mixing meat and dairy foods, the idea of eating a heretofore taboo combination was intriguing to me, so naturally I ordered one. As I gingerly picked it up to take a bite, my stomach did a flip-flop, and because the taboo was so strongly ingrained, I was unable to enjoy the combination. Modern food science had produced a fake cheese, which allowed the Orthodox consumer to experience a new taste sensation. Apparently others may have felt as I did, for this dish was soon discontinued from the restaurant's menu.

Cutting-edge science may now afford us the opportunity to create other new foods for our kosher tables, and some of those foods may revolutionize the choices available for the kosher consumer. Genetic engineering allows scientists to manipulate the traits of many different types of organisms, including plants and animals, producing organisms with new and novel traits that have not previously been encountered within a particular species. Plants and animals that have been genetically altered in this way are called "transgenic."

The first transgenic animal ever produced with novel traits was a laboratory mouse whose fertilized egg was injected with a rat gene for growth hormone.[1] When the gene was turned on, it enabled the mouse to grow to about twice the size of a normal mouse. The mouse was dubbed "Mighty Mouse." Other more recent experiments on animals involve the genetic engineering of farm animals. Transgenic goats are produced by injecting goat embryos with specific genes. Some of these animals produce pharmaceutical drugs, like t-PA (tissue plasminogen activator), which is used to treat heart attacks. Scientists can harvest those products from goat's milk and purify them to produce relatively low-cost drugs.

When transgenic plants and animals are produced, a specific gene is first identified and isolated from living tissue. The source for the gene can be plant, animal, fungus, or single-celled organisms, such as bacteria. DNA from the donor species is purified chemically, cut into pieces, and copied millions of times in the lab (either in a bacterial cell, or by chemical methods). The resulting purified, amplified gene is transferred to a recipient species to alter the genetic program.

The donor species provides the DNA sequences, or genetic information, that determines a particular trait. The recipient or host species must accept the DNA and link the genetic information to one of its own chromosomes so it becomes part of its own genetic program. In order for the process to work, the genes must be intact and be able to be activated or turned on in the recipient. Creating a transgenic plant or animal is therefore not a simple task. Since our understanding of gene action is far from complete, frequently the procedure is unsuccessful. Because of the difficulties involved, and the expense and time needed to succeed with such projects, genetic engineering is not done for frivolous reasons. There must be a very compelling reason to try to change the genes of an organism, and hence its traits. The justification for such experiments include clinical, environmental, and nutritional applications, and, perhaps most compelling in a capitalist society, economic gain.

Although there are many potential technical difficulties, and the monetary investment may initially be high, genetic engineering to produce transgenic plants and animals has become quite important in the commercial world. New genetically engineered species of plants are already being sold in supermarkets (see chapter 12). A number of species of animals have been successfully altered by bioengineering. Food and Drug Administration (FDA) regulations do not require the labeling of foods that have been genetically altered, so consumers unknowingly have been eating these genetically modified—or GM foods—for the last several years. Since we are already partaking of the products of genetic engineering, an important question for the Jewish consumer is: How do these new genetically engineered foods affect the kosher dinner table?

How Are Transgenic Animals Produced?

The production of a transgenic animal involves the merging of reproductive science and molecular techniques. Since in vitro fertilization (IVF) was developed, scientists have learned a great deal about eggs from many mammalian species, including how to grow them in petri dishes and how to manipulate them. Eggs from mice, sheep, goats, pigs, and humans are exceedingly small, even smaller than the period at the end of this sentence. Thus, they can barely be seen—even by a trained technician—without a microscope. So all manipulation of eggs must be done under a microscope with specially designed microscopic tools.

Pronuclear injection is one method used to make transgenic animals. It involves fertilizing eggs in a petri dish, then individually injecting DNA directly into their pronuclei—the genetic compartments of the newly fertilized eggs. Mammalian eggs that have just been fertilized have two pronuclei, one containing the genes from the mother and the other holding the sperm's genes, that is, genes from the father. DNA from just about any source can be introduced into the host species by injecting it into the pronuclei. The free ends of the injected DNA strand become linked to the embryo's DNA and will remain part of its chromosome. The genetically modified embryos are implanted into a surrogate mother of the same species. This procedure is not very efficient; of the offspring born, only 1 to 5 percent are transgenic, having linked the DNA to their own. And only a small percentage of those transgenic animals will carry active genes, that is, genes that actually work in the animal. "Although pronuclear injection is widely used, the low efficiency of transgenic production and unpredictable level of protein expression have made the system costly and inefficient," reports Rathin Das.[2]

A more efficient, albeit more complicated method, involves nuclear transfer, also known as cloning. In this procedure, a fertilized embryo of the species to be modified is grown until it is a ball of cells, or a blastocyst. Some cells are removed from that embryo and grown in a petri dish. Compared to microinjecting eggs (which is a laborious and tedious task), it is easier and more efficient to coax DNA into those cells by a variety of techniques. Techniques include the use of an electrical pulse to draw the DNA through the cell membrane, the use of a virus to carry DNA into the cell, or the microinjection of DNA into each individual cell. This type of genetic alteration is much more efficient because many cells can be processed at once, and the cells that have not picked up active genes can be discarded. Only those that have successfully linked the genes to their own chromosomes and are genetically modified (GM) are chosen for the next step. In the next step, eggs of the same species are collected, and their own nuclei are removed. GM cells are fused with the "empty" eggs. The nucleus of each GM cell enters an egg and will then take over and direct the development of the embryo (utilizing the modified genes). The embryo is implanted into a surrogate mother of the same species, and grows into a transgenic, genetically modified animal.

The problem with nuclear transfer, a form of cloning, is that many of the offspring produced by this method die before or soon after birth. There are many developmental problems found in these animals, so that the success rate of this technique is low. With pig cloning, for instance, only 1 to 2 percent of the embryos are born alive. Scientists still do not understand the underlying reason for the high death rates of these cloned animals.[3]

Despite these problems, it is clear that production of transgenic animals makes sense from an economic and scientific standpoint. When the experiment is successful, it can produce an animal that makes a pharmaceutical drug. That animal acts like a drug factory, and each animal becomes extremely valuable since it can outproduce traditional methods of drug preparation. Das comments, "Transgenic animals provide an alternative and potentially more economical source of production of protein therapeutics compared to the traditional cell culture and fermentation systems."[4]

"Pharm" Animals: A Barnyard of Genetically Modified Creatures

In a recent article entitled "If It Walks and Moos Like a Cow, It's a Pharmaceutical Factory," Carol Yoon[5] enumerates the array of novel animals that have already been produced by these methods. Many of these mammals secrete pharmaceutical products directly into their own milk, so recovery of the drug is easy and painless for the animal. The animals are milked, the milk is collected, and the product may be chemically purified from the milk.

These procedures have produced some unusual creatures, referred to by some as "pharm" animals. There are cows that make lacto-ferrin, a protein used to treat infections; goats that produce anti-thrombin III, used to prevent clotting in people; and sheep that make milk containing alpha-antitrypsin (AAT), a drug used to treat AAT deficiency disease, a "leading genetic killer of adults in the United States."[6] Growth hormone, hemoglobin (for anemia), and Factor VIII (for hemophilia) are a few other examples of proteins produced in this manner. My personal favorite is a goat that produces spider silk in its milk. The silk protein can be purified from the milk to make sur-gical sutures or strong, lightweight, bulletproof vests. Perhaps the next novel animal will be an all-in-one goat whose wool *and* silk fi-bers can be used to manufacture the currently popular and very ex-pensive fabric called pashmina.

When chickens are genetically modified, since they do not produce milk, the results are found in their blood or their eggs. Some of these birds lay eggs containing pharmaceuticals. The drugs can be purified, or consumed directly with the egg (perhaps redefining the notion of a healthy breakfast.)

For lactose intolerant people, instead of having to take lactase (an enzyme that breaks down lactose) separately, how about a milk that already contains the enzyme? Researchers have produced lactase in mouse milk.[7] (And, not surprisingly, there is not much of a market for it.) They are currently in the process of scaling up from mice to livestock.[8] Another unique creature already produced is the geneti-cally modified pig that makes a cleaner stool. These animals produce an enzyme that reduces phosphorus in their manure, which can lead to a reduction in phosphorus pollution of soil and groundwater.

Genzyme Transgenics, a company located in Framingham, Massa-chusetts, claims it has produced more than sixty therapeutic proteins in the milk of a variety of transgenic animals. Many of these products are being tested for eventual use in patients, although only a handful of them have progressed to the point where human trials are being conducted.[9]

The global market for transgenic animals was $13 billion in 1999. The industry can be very lucrative because pharmaceuticals are prescribed in microgram and milligram quantities (1 millionth and 1 thousandth of a gram, respectively). Transgenic animals can produce as much as 20 grams per year (each rabbit), 250 grams per year (each chicken), a kilogram per year (each pig), and more than 20 kilograms per year (per each sheep, each goat, and especially, the mother of all "animal factories," the cow). In order to understand the potential impact these animals will have, consider the impact of just one product: Malaria antigen, MSP-1, produced in the milk of transgenic mice. This protein can be used to produce malaria vaccines capable of saving the lives of 2 million people who die each year in tropical countries.[10]

Salmon have been genetically engineered to produce transgenic fish that grow twice as fast as normal salmon. This allows fish farms to produce market-size fish in eighteen months, versus almost three years. These fish, and others like them, have been reprogrammed to produce growth hormone all year round, instead of only in the warmer months. Animals such as these have been touted as one potential response to global hunger. Fish that grow faster, needing fewer resources, could cost less pound for pound and feed more people. However, environmental advocates are concerned that if fish with unique traits escape into the wild, the impact on the environment would be unpredictable, and possibly disastrous. For instance, since the genetically modified male fish grow faster, they may be preferred mating partners for wild female fish. But their genetic modifications— if released into the wild—could lead to the so-called the "Trojan gene effect": the altered males with traits that make them attractive to females could outcompete wild males for mates, but they may carry genes that lead to weaker offspring who might not survive in the wild. "Populations of wild fish could, in theory, be wiped out by mating with certain kinds of genetically engineered fish, should they escape," cautions Yoon.[11]

Is It "Kosher" to Genetically Modify Animals?

The prospect of genetically modified fish, pigs, and other species leads to speculation that genetic engineering could alter previously unkosher species and make them kosher. Alternatively, kosher species might be modified by introducing genes from unkosher species. Would that render those animals *treife,* or unkosher? (A diseased, defective, or abnormal animal is considered *treife*.) Is it permissible under Jewish law to alter species and produce barnyards and ponds filled with GM animals? If so, how would the alterations be viewed in terms of kashrut (kosher status)?

According to the laws of kashrut, animals are judged kosher if they exhibit split hooves and chew the cud. The pig has a split hoof, but does not chew its cud. The camel is a ruminant, but does not have a split hoof. If those species could be modified to exhibit both traits of kosher animals, would the modified animals be accepted as kosher? According to biblical stipulations, in order for fish to be considered kosher, they must have fins and scales. (Deut. 14:9) Under these definitions, catfish and eels are not considered kosher. If it were possible to genetically engineer those species so that their traits were altered, and they became scaled fish, would they be considered kosher?

"Whether or not there is a specific midrashic reference to a pig which chews the cud, it would appear that an animal which has split hoofs and which also chews it cud is *ipso facto* kosher," states Rabbi J. David Bleich.[12] Bleich is referring to an unusual animal, the babirusa, which at one point had been suggested to be a version of kosher swine. The babirusa is a species of swine, native to Indonesia, that has split hooves (like a pig) but was also reported to chew its cud. It turned out, after careful study of this animal, that it is not a true ruminant. Thus, the question of the babirusa being a kosher pig became a moot one. However, in his analysis of this species, Bleich touches upon some important issues relevant to kashrut of genetically modified animals.

For instance, Bleich explains that animals that chew the cud also appear to have teeth that are compatible with grazing. They lack incisors and canines used by carnivores to tear meat, and instead have flattened surfaces on the upper jaw that are more useful for grinding vegetation. Bleich states, "Absence of incisors and canines is itself evidence that the animal is a cud-chewing ruminant."[13] In addition, whether an animal has horns can be an important sign. "The presence of any type of horn is indicative of the fact that the animal is a member of a kosher species," reveals Bleich. "The presence of horns is sufficient to exclude the possibility that an animal may be a pig."[14] So in order to create a novel form of kosher pig, it appears that the modified species should not only chew the cud but also have teeth characteristic of a ruminant animal and, at some point in its development, grow horns. The production of this type of animal would involve introducing a complex set of traits, necessitating the introduction and expression of numerous genes. It will certainly be difficult, if not impossible, to produce such complex alterations in a pig and maintain its viability.

Kilayim: *Is Genetic Engineering Considered a Forbidden Mixture?*

One of the major arguments against genetic engineering is that material from two distinctly different species is mixed together to form a new animal or plant. Leviticus (19:19) states: "You shall keep my statutes. You shall not let your cattle gender with a diverse kind; you shall not sow your field with mingled seed; neither shall a garment mingled of linen and woolen come upon you." This passage is the basis for the prohibition of *kilayim,* or forbidden mixtures (translated by some as forbidden junction).

There are two main categories of commandments in the Torah: *mishpatim* and *khukim. Mishpatim,* or ordinances, are "commandments, the observance of which would have been deducible even if the Bible had not enjoined them." *Khukim* (singular form, *khok*) are the statutes or "commandments which would not have been logically derived unless specifically commanded."[15] The prohibition involving *kilayim* is considered a *khok,* because of the preface, "You shall keep my statutes," and also since logical reasons for these laws may not be readily discernable. A *khok* is a commandment that proves a Jew's faith, since human logic cannot explain it.

There is a tractate of the Mishnah entitled *Kilayim* that only addresses issues related to these statutes. For instance, it explains which animals should not be bred together, which seeds cannot be sown together in the same field, and what types of garments are forbidden when woven from two different threads. The combination of sheep's wool and linen, when woven together in a garment, is prohibited. This combination is called *shaatnez,* which is a contraction of the Hebrew words for "comb, spin and weave."[16] Orthodox Jews still stringently observe this commandment and do not buy clothes made from this mixture of materials. Observant men who buy new suits may have their garments laboratory tested to ensure that no linen is found in a woolen suit.

Regarding admixture of animal species, the Mishnah is very specific regarding which combinations are permitted and which are prohibited. For instance, it is prohibited to mate a horse with a donkey. The product of such a mating, a mule, is a useful work animal but is sterile. The Mishnah stipulates that some other animals, although similar in appearance, also are forbidden mixtures: "The wolf and the dog, the village dog and the fox, goats and deer, mountain-goats and ewes, the horse and the mule, the mule and the ass, the ass and the wild ass, although they resemble one another, are forbidden junction one with the other."[17]

It is interesting to note, however, that once the offspring of a forbidden mixture have been made, Jews are permitted to use them. "Diverse kinds among cattle are permitted to be reared and to be maintained, and are only prohibited from being bred; diverse kinds of cattle are forbidden to be mated one with another."[18]

Do these statutes apply to the question of genetic engineering? Most of the Mishnah has been expounded upon by the Gemara. However, in this case, there is no Babylonian Talmud corresponding to Mishnah *Kilayim*. The Jerusalem Talmud does address issues in *Kilayim*. Rabbi Shabtai Rappaport, an Orthodox rabbi from Efrat, Israel, explains that "the Jerusalem Talmud (*Kilayim*, ch. 1) interprets 'my statutes' as the statutes of creation, rather than legal statutes."[19]

In other words, expounds Conservative Rabbi Shammai Engelmayer, "The prohibition in Leviticus begins with the words: 'Eht chukotai tishmoru,' 'You shall keep My statutes.' The word for statute, chok, shares the same root as chakok, which means to engrave or to carve. Thus, the Jerusalem Talmud reasons, it is forbidden to change something that was 'carved out' by the Creator at Creation."[20]

On the basis of this interpretation, some have suggested that mixing genes between species constitutes changing the order of creation and, by virtue of the laws of *kilayim*, would be prohibited. The thirteenth-century commentator Nachmanides, also known as Ramban, wrote: "And the reason for *kilayim* is that God created species in the world of all the living creatures, the plants and the animals, and gave them reproductive powers so that they would maintain the species for as long as He wished them to exist . . . and He commanded them to reproduce their own kind, and they shall not change as long as the world exists . . . and whoever breeds two different kinds together is adulterating the works of creation; it is as if he thinks that

God did not complete the world, and he wants to help creation by adding creatures."[21]

A Reform responsum discusses Ramban's view on *kilayim*. "Ramban . . . stated that human beings should not change nature as that would imply imperfection in God's creation. That medieval view was found frequently in church literature. It has not been followed by Jewish thinkers. Jewish law said nothing about changing the characteristics of a particular species of breed. Throughout the centuries every effort was made to assist nature and to produce animals suited to specific purposes as well as plants which would yield abundantly."[22]

And, indeed, other Jewish thinkers also differ with Ramban on his interpretation of *kilayim*. Rappaport recommends reexamining Ramban's view. "It is an accepted ruling, as noted by the Chazon Ish . . . one of the leading Torah authorities of the twentieth century, that artificial insemination to generate a hybrid is indeed permitted because the prohibition to 'let your cattle gender with a diverse kind' applies only to sexual contact between living animals."[23] Genetic engineering, involving the introduction of DNA from one organism into another certainly does not involve sexual contact. If artificial insemination is acceptable, then DNA technology, where only a portion of the genes are inserted into an unrelated species, should certainly be permissible.

Rappaport maintains that since God gave man the power to develop technologies, man is permitted to use those technologies. Rappaport states, "If artificial insemination is considered a human technology which is not governed by the 'ordinances of heaven,' then this formula also applies to genetic manipulation. The halachic viewpoint is that the more advanced the technology, the more reason there is to permit it. Any consideration of an organism being an unchangeable entity has no basis in halachah."[24]

On the other hand, Engelmayer, who is more reluctant to accept genetic engineering, asks whether the nature of creation itself may be manipulated. "In Genesis 1, the language is very specific. Everything was created 'after its kind . . . , and God saw that it was good.' Clearly, the very first chapter of the Torah states that God built into His creation pointed differences between one thing and another—and He liked it that way. Dare we fiddle with what God wrought?"[25]

A leading halakhic authority in Israel, Rabbi Shlomo Zalman Auerbach, supported the use of genetic engineering; he did not believe that it constituted *kilayim*. Dr. Avraham Steinberg reports Auerbach's views that "genetic material may be transferred from one species to another by means of bacteria or viruses, and the resultant species is not considered to be *kilayim* (forbidden mixtures)."[26]

Even if one does not accept artificial insemination between species, because it may violate *kilayim*, there are additional, even more compelling reasons to accept genetic engineering. The laws of *kilayim* are incredibly specific, such that the smallest divergence in a situation may render the mixture permitted. For instance, although it is forbidden to weave wool and linen together, if camel's hair and sheep's wool are spun together, and the yarn contains a greater proportion of camel's hair than sheep's wool, it is permitted to weave it together with linen. In addition, if hemp (a fiber used to make rope and canvas) and flax (the raw material from which linen is derived) are spun together, and the yarn contains a greater proportion of hemp than flax, then it is permissible to weave it together with wool.[27] A commentary on this Mishnah explains, "The well-known maxim applies, a minority becomes annulled in a majority, or a major disannuls a minor quantity, or the lesser is canceled by the larger."[28]

Thus, although it is prohibited to mate two different species, which involves combining approximately equal parts of the genetic material to create a new species, it could be inferred from the Mishnah that, if only a small portion of genes from one species is added to a different species, the prohibition is overridden. The minority of DNA sequences of the donor species would be annulled in the majority of DNA found in the host. As long as the quantity of DNA of one species exceeds the other, the prohibition should not apply.

Furthermore, it is important to keep in mind, that even the products of true *kilayim* are permitted for use. Rabbi Avram Reisner writes, "Use of the product of the hybridization is affirmatively permitted. In fact, many hybrids are presently on the market, both hybrids of different strains of the same type of plant, which would not be kilayim, and those of separate species, which would be considered kilayim, the product of agricultural and animal husbandry techniques honed before the advent of genetic engineering. No such product is banned. Indeed, this is not even a modern leniency, having its earliest source in the Tosefta."[29]

As the *Shulhan Arukh* states, "It is forbidden to maintain *kilayim*, but the fruit produced thereby is permitted even to the one who transgressed and produced the hybrid. It is permitted to take a branch from the hybrid and plant it elsewhere."[30] Thus, if genetic engineering is one day judged by rabbis to be a form of *kilayim,* although Jews would not be permitted to produce such species, they would be able to benefit from the new life forms. And once the new life form has been produced, it appears that Jews would also be permitted to continue to breed the genetically engineered species, since the *Shulhan Arukh* permits the hybrid to be planted elsewhere.

Tza'ar baalei khayim: *The Suffering of Animals*

Another issue that is critical in evaluating the permissibility of genetic engineering techniques is that of *tza'ar baalei khayim,* or causing the suffering of animals. Throughout Scripture there are admonitions regarding the rights of animals, and an emphasis on the need to treat animals with kindness and respect. "Animals could be used by man as long as they were treated kindly. It is prohibited to consume a limb from a living animal. . . . An animal which was threshing may not be muzzled; it must be permitted to eat as freely as a human being. . . . Furthermore, one should not consider acquiring an animal unless one has the means to feed it . . . and a person should then feed his animals before feeding himself. . . ."[31]

In Proverbs (12:10) an overriding principle regarding animal welfare stated, "A righteous man regardeth the life of his beast; but the tender mercies of the wicked are cruel." God's concern for His creatures is noted in, "And His tender mercies are over all His works" (Ps. 145:9) and "Man and beast Thou preservest, O Lord" (Ps. 36:7).

The obligation to show kindness to animals includes the lightening of a beast's burden (Exod. 23:5), the prohibition of killing a mother with its young (Lev. 22:28; Deut. 22:6), and the requirement that "on the seventh day thou shalt rest that thine ox and thine ass may have rest" (Exod. 23:12). In addition it is prohibited to castrate animals (Lev. 22:24).

Maimonides (also known as the Rambam) wrote in his *Guide to the Perplexed* that it is "allowed, even commanded, to kill animals; we are permitted to use them according to our pleasure." However, "There is a rule laid down by our Sages that it is directly prohibited in the Law to cause pain to an animal." The motive for those commandments is to improve human behavior. "The object of this rule is to make us perfect; that we should not assume cruel habits; and that we should not uselessly cause pain to others; that on the contrary, we should be prepared to show pity and mercy to all living creatures, except when necessity demands the contrary."[32]

Indeed, the ancient view that cruelty to animals can lead to callousness and subsequent cruelty to people has been confirmed by modern studies of the human psyche. In fact, one of the diagnostic criteria for conduct disorders is exhibiting physical cruelty to animals.[33] It has been observed that youngsters who practice animal abuse may go on to abuse, harm or kill people.[34] An FBI study in the late 1970s revealed that 46 percent of convicted multiple murderers had, as teenagers, "deliberately caused severe or repeated suffering to animals."[35]

As noted by Maimonides (above), Jewish law does sanction the use of animals, even that use which causes animal suffering, if the need is justified. However, the extent of the suffering should always be kept to a minimum. For instance, the slaughter of animals for food is permitted. However, Bleich explains that ritual slaughter should be done in a way that will cause the least suffering. He cites the *Sefer haHinukh*, "with regard to the reason for slaughter at the throat with an examined knife that it is in order that we not cause pain to animals more than is necessary, for the Torah has permitted them to man by virtue of his stature to sustain himself and for all his needs, but not to inflict pain upon them purposelessly."[36]

Tza'ar baalei khayim is permitted for the purposes of developing new treatments or cures to alleviate the suffering of humans. Thus, the use of animals in medical experimentation would be sanctioned by Jewish law. (Note that human life is considered so precious that Jewish law sanctions the violation of almost all of the commandments—including the Sabbath and the laws of kashrut—in order to preserve it.) However, there is a controversy concerning whether *tza'ar baalei khayim* is permitted for financial reasons. Bleich cites one view (the Ritva, *Shabbat* 154b), that *tza'ar baalei khayim* is permitted for financial profit. However, according to Bleich, another view (from *Teshuvot Avodat haGershuni*) maintains that it "cannot be sanctioned for purposes of realizing 'a small profit.'" The *Tosafot,* a commentary on the Gemara, permits it even for "non-life-threatening maladies," but not for monetary gain.[37]

The Reform Jewish view is similar to Orthodox rulings. The New American Reform Responsum on Jewish Involvement in Genetic Engineering states, "When dealing with experimental animals we should be quite certain that they are not subjected to pain or used for frivolous reason as for example cosmetic experimentation."[38]

Bleich, writing from the Orthodox perspective, concludes that

Jewish law clearly forbids any act which causes pain or discomfort to an animal unless such an act is designed to satisfy a legitimate human need. Although many authorities maintain that it is not forbidden to engage in activities which cause pain to animals in situations in which such practices yield financial benefits, there is significant authority for the position that animal pain may be sanctioned only for medical purposes, including direct therapeutic benefit, medical experimentation of potential therapeutic value and the training of medical personnel.[39]

Any genetic engineering experiment that would result in pain or suffering of animals would have to be scrutinized in order to determine how to keep the suffering of animals to a minimum. The motive for these experiments would determine whether Jewish law permits such use of animals. For instance, if genetic engineering produced an animal model useful for cancer research (e.g., the so-called Oncomouse), then this is clearly designed to assist researchers in the understanding of a grave human disease and would be permitted by Jewish law. As Rabbi Nisson Shulman reports, "There is no objection to new products such as the 'Oncomouse' for use in laboratory experiments in order to cure illness, since man is given dominion over this world which is created for his use."[40] Likewise, if the experiment involved the production of goats that make important pharmaceutical drugs, the motive is to alleviate human suffering and disease, and there should be no halakhic problem with using animals in that way.

On the other hand, if the goal is to make an animal that produces a new fiber, and the fiber's application would be only commercial, and not potentially lifesaving, then the pain and suffering inflicted on the animal may make that application unacceptable. Note that for one genetically modified animal mentioned above, a goat that produces silk protein in its milk, the product of that genetically modified goat has applications in the production of sutures and bulletproof vests and can potentially preserve human life. It probably is no accident that those are the applications emphasized by the researchers. It is certainly possible to imagine other uses for such fibers that would have only commercial and not life-saving applications (e.g., production of designer clothes, pocketbooks, home furnishings, etc.).

Does *"Heal, He Shall Heal"* Give Permission to Engineer Animals?

The application of genetic engineering for medical treatments or cures is permitted by even the most stringent rabbinic interpretations of the technique. Reisner states, "Judaism's emphasis on healing individuals who are sick is likely to override any combination of concerns that might otherwise impact the technique."[41]

Nachmanides supported the use of scientific knowledge for the benefit of mankind.[42] The Human Genome Project, the goal of which is to determine the complete DNA sequence of the human, would by this measure be a biblically mandated pursuit. The knowledge of DNA sequences will enable the understanding of the disease process, and ultimately the alleviation of human suffering.

"Knowledge and its pursuit are legitimate activities for human beings and not considered an encroachment upon Divine prerogatives," declares Rosner. "Thus, therapeutic genetic engineering and gene therapy that may result from the knowledge derived from the genome project is not a biblical violation of undermining G-d's Creation of the world by manipulating nature."[43]

Pikuakh nefesh (the preservation of human life) is cited by Rabbi Walter Jacobs as a principal reason to permit these techniques. "We may be ready to accept genetic changes made for medical purposes and experimentation as pikuah nefesh in an overriding consideration . . . ," Jacobs states. "Human life must be saved if it is at all possible and even some pain to animals is permitted for this purpose. Economic reasons, however, could not justify such a course of action. These should always be reviewed carefully."[44]

Can Nonkosher Species Be Rendered Kosher?

Genetic engineering of animals could, presumably, alter the traits of a nonkosher animal such that it exhibits the traits of a kosher animal. It could change scaleless fish to fish with scales, and might be used to alter traits of pigs, horses, rabbits, camels, and the like, so that they exhibit traits of kosher animals. If you could change the traits in this manner, would it be possible to produce a kosher species from a nonkosher one?

Bleich's approach to understanding the nature of the babirusa (a pig that reportedly chews the cud), once again illuminates issues regarding genetically engineered species. "Assuming that the babirusa manifests the requisite criteria of a kosher animal," states Bleich, "the fact that it resembles a pig in appearance and in taste is not sufficient ground for banning its consumption as kosher meat."[45] Thus, even though it looks like pork and tastes like pork, if it is from an animal that fulfills the criteria of kashrut, it is permissible to eat it. (On the other hand, snake meat, which reportedly looks like and tastes like chicken, is decidedly not kosher!)

However, there are some important stipulations with regard to kosher animals that could dramatically affect any decision on genetically modified species. First is the issue of *yoze*, which means emanating or born from. "Codifying a principle laid down in the Mishna . . . [the *Shulhan Arukh*] rules that the offspring of an unclean mother is non-kosher even if the animal itself manifests all the characteristics of a kosher animal," explains Bleich. "The principle is that of "yoze," "i.e., anything which 'emerges' from, or is produced by, an unclean animal is itself not kosher. It is on the basis of this principle that, for example, the milk of nonkosher animals is forbidden."[46]

Rabbi Hayyim ha-Levi Soloveichik is cited by Bleich for his view that species is determined by parentage. In other words, an animal is classified regarding its species on the basis of its mother. "It is, then, maternal identity which is transmitted to progeny and which determines the species to which the offspring belong for purposes of halakhic classification. On the basis of either analysis, the offspring of a non-kosher animal is not kosher even if, as the result of genetic mutation, it manifests the criteria of a kosher animal."[47]

The Aish Hatorah webpage contains interesting responsa to modern-day topics. That group approaches questions from a traditionally observant perspective. According to their responsum on genetic engineering, "Gene therapy has raised some interesting questions in the laws of kosher food. The Talmud (*Bechorot* 5) discusses a case where an animal that appears like a cow (a kosher species) is born to a camel (a non-kosher species). Is this new animal kosher or not? The Talmud concludes: Anything that comes out of a non-kosher animal is considered not kosher." However, the responsum declines to draw a conclusion with regard to genetically engineered organisms.[48]

There is at least one other significant problem with a "kosher pig."

Although the Bible clearly states, "And every beast that parteth the hoof, and hath the hoof wholly cloven in two, and cheweth the cud, among the beasts, that ye may eat," the Talmud[49] stipulates that the only animals considered kosher are those listed in Deuteronomy 14:4–5. According to that viewpoint, all kosher animals must have arisen from the ten species listed: "the ox, sheep and goat; the hart, deer and the yachmur, the akko, dishon, the teo and the zamer." The first five species are known, but the last five, according to biblical commentator Abraham Ibn Ezra, "can only be established by tradition." The Soncino translation offers the last five species as "the roebuck, and the wild goat, and the pygarg, and the antelope, and the mountain-sheep."[50] But the ArtScroll Stone Edition[51] avoids translating those five species into English, since the exact identity of these species is open to some speculation. Rabbi Binyomin Forst lists kosher animals as "cows, goats, sheep, deer, bison, gazelle, antelope, ibex, addax and giraffe."[52]

However this list of kosher animals is translated, we do know that the following species are not kosher: "the camel, and the hare and the rock-badger, because they chew the cud but part not the hoof . . . and the swine, because he parteth the hoof but cheweth not the cud . . . of their flesh ye shall not eat and their carcasses ye shall not touch." (Deut. 14: 7–8) Those biblical statements make it very clear that swine are not kosher, and are not of the ten kosher species.

Bleich also reports that "according to a number of latter-day authorities, it is forbidden to eat the meat of any hitherto unknown species even if it possesses the characteristics of a kosher animal and does not in any way resemble a non-kosher species. Hokhomat Adam 36:1 declares '. . . we eat only [those animals] with regard to which we have received a tradition from our father.'"[53]

Given these specifications, a genetically engineered pig, which is the offspring of a mother pig, even if it is transformed into a cud-chewing, split-hoofed creature, will not be declared kosher by most authorities.

Can Kosher Species Be Rendered Nonkosher?

Genetic engineering may involve the introduction of genes from nonkosher species into kosher species. In the production of genetically engineered goats, cows, fish, or other animals, DNA from other species

may be used. For instance, human DNA can induce genetically modified animals to produce human proteins for use in medical therapy.
Silkworm genes are transferred to goats to make transgenic goats
that produce silk protein. Would that render those kosher animals
nonkosher? Reisner addresses the question by considering basic ten-
ets of the laws of kashrut. "The primary rule is that they are forbidden *binotayn ta'am*—when they impart a flavor to the resultant
product. This is estimated, as a matter of law, at one part in sixty,
such that a lesser admixture is permitted, a greater admixture forbidden."[54] For instance, if one is cooking a large pot of meat broth, and
a droplet of milk accidentally falls in, it would not necessarily render
the whole pot of soup nonkosher, even though meat and milk should
not be mixed and eaten together. The concept of *notayn ta'am*—impart a flavor—is critical, since any nonkosher ingredient that changes
the nature of the food would be prohibited even if present in less
than one sixtieth portion. For instance, food stabilizers and concentrated food flavoring agents can affect the nature of the food in very
small amounts, and must be from kosher sources. In addition, if one
intentionally mixed in a nonkosher ingredient (or put milk into meat,
or meat into milk), even if it were less that one sixtieth part, the
product would be forbidden.

Reisner points out that, in the case of genetic engineering, the DNA
that is used is always significantly less than one part in sixty. But the
genetic engineering is usually done to change the nature of the recipient in a significant way. In addition, it is administered intentionally; it
is deliberately meant to change the nature of the organism.[55]

"Halakhah had to distinguish between that which is counted and
that which is nullified, that which is perceptible and that which is
not," explains Reisner. " In doing so, Jewish law in the modern period
has settled on the rule of thumb that microscopic items, not visible to
the naked eye, are discounted altogether in determining Jewish law."
This statement may be key in determining how genetic engineering affects the kashrut of animals or plants. Reisner based this statement on
a ruling published by Rabbi Yechiel Epstein, in his seminal work
Arokh HaShulhan, written in the 1890s.[56] By the end of the nine-
teenth century, microscopes were already being used routinely to
study microscopic cells, which comprise all living things, and to identify microorganisms that cause disease.[57] Epstein reasoned that microscopic life cannot be considered in the determination of whether

something is kosher or nonkosher. The *Arokh HaShulhan* states that because of the myriad microscopic creatures in our water and in our air, "were we to consider microscopic life forms we would be unable to drink the water or breathe the air."[58]

Another major principle with regard to the kashrut of genetically modified animals and plants involves inedible foods, called *ayno ra'ooey l'akhila*. "Any forbidden food, whether of Biblical or Rabbinic origin, which becomes unfit to eat, loses its status as a forbidden food," states Forst. He adds, "Non-kosher food which has spoiled and become completely inedible *and* mixed into kosher food may be eaten even if the non-kosher food is a majority of the mixture, providing that the non-kosher food cannot be separated and removed."[59] In the category of *ayno ra'ooey l'akhila*, toothpaste is one example of an essentially inedible substance which may be used, and even unintentionally swallowed, regardless of whether it has some nonkosher ingredients.

When DNA is extracted from animals (or plants), chemicals are added that make the flesh inedible. The purified DNA itself has no nutritional value of any significance, and is not considered a food. Placing DNA from a nonkosher source into a kosher animal or plant would be analogous to mixing inedible nonkosher food with kosher food. Even if the inedible food is in the majority, the mixture would be permissible. The protocols used in genetic engineering only require submicroscopic quantities of DNA, which should not be counted as halakhically significant, since the DNA is invisible to the naked eye, and, like inedible foods, should not transform kosher foods into nonkosher foods. "Judaism does not regard genes as 'food,'" states Rabbi Nisson Shulman in his volume of responsa from the Office of the Chief Rabbi of Great Britain. "So that the introduction of a gene or genetic material from non-kosher animals would not render the recipient non-kosher."[60]

The questions involved in the production of kosher cheese are very relevant to this discussion. For thousands of years, kosher cheese has been produced by adding small amounts of the active ingredient rennet to raw milk. Although rennet is derived from animal stomachs, cheeses are not rendered nonkosher by virtue of these small amounts of animal products mixing with milk. Rennet contains the enzyme rennin, which curdles the milk, changing its consistency from liquid to solid. (In nature the curdling effected by that enzyme helps to slow down the movement of milk from the stomach to small intestine, so that it can be more completely digested. In fact, when a infant spits up his milk, it can have a cheesy appearance, as it has already been curdled by the infant's own stomach enzymes.)

One might imagine that cheese could never be kosher since an animal product is being mixed with milk. "Rennet has lost its identity as food, because it's not edible. It has become a chemical catalyst," explains Rabbi Tzvee Zahavy. "Anything that is not edible for a dog is the threshold. Once it loses its identity as food, it is not governed by the laws of kosher food."[61]

In production of transgenic plants or animals, there is no direct transfer of "ingredients" from one species to another during the genetic engineering process. DNA is isolated in the lab, copied by bacteria or by chemical methods, and then inserted into the recipient cell. "DNA is a chemical and wouldn't be edible for a dog," reasons Zahavy. "It is absolutely not governed by the halakha of food."[62]

The latest development in the kosher cheese industry is the use of genetically engineered chymosin (microbial rennet). "Industrial scientists now have the ability to turn microorganisms into microscopical chemical factories," explains Dr. Judith Leff, who served as scientific consultant for several kashrut organizations. "They churn out their products in huge multi-storied types of vats called fermentors. Thus, bacteria, fungi, or yeast can be harnessed to produce a whole array of food ingredients. . . ."[63] Although some microorganisms make their own version of chymosin, which can produce cheese, the quality of cheese made with microbial rennet was not adequate. Genetic engineering solved that problem. Cow genes were introduced into microorganisms, and the genetically modified cells were altered to produce calf rennet identical to that made in calf stomachs. According to a recent report, Chymogen, a genetically engineered version of chymosin (rennet), is used to make sixty percent of all hard cheese products.[64]

"Kashrut agencies have permitted the use of genetically engineered chymosin (microbial rennet) in the production of cheese," reports Reisner. "Similarly, here, genetic transfer happens at a submicroscopic level which the halakhah is hard-pressed to consider." Reisner concludes, "Thus, the rules of kashrut, the rules of admixtures, simply fail to address the nature of transgenic creations. Absent a reason to declare the new product unkosher, it would appear to be fit for consumption."[65]

DNA *as Transplanted Tissue*

When DNA is transferred from one species to another, the process could also be considered analogous to a transplant, since the DNA is introduced into a living organism and becomes an active part of the organism. It is not just an ingredient in a food recipe.

Dr. Fred Rosner discusses the case of a cornea transplant. He cites Rabbi Jekutiel Judah Greenwald, who maintained that an "engrafted or transplanted cornea becomes nullified on the recipient."[66] This conclusion concerning transplanted tissue is grounded in part on the talmudic passage: "If he grafted a young shoot on an old stem, the young shoot is annulled by the old stem and the law of Orlah does not apply to it." (*Orlah* refers to the prohibition of using the fruit from a tree for the first three years after it is planted [Lev. 19:23]). The young shoot does not retain its status; it acquires the status of the old tree.[67]

Rabbi Azriel Rosenfeld who wrote one of the first halakhic analyses of genetic engineering also considered the process akin to transplantation. Citing Rabbi Eliezer Deutsch, regarding an ovary transplant from one woman into another, "Once the donor's sex organs are in the recipient's body, they become part of that body. . . ." In fact, he revealed that a child conceived from an egg from a transplanted ovary has no relationship to the donor. Rosenfeld concluded, "This ruling too would surely apply in a gene transplant case."[68]

Shulman supports the proposal that the "genetic material is considered part of the host. It is not regarded as part of the donor organism at all. . . ." He goes on to explain, "A permitted food would therefore not become forbidden simply because of gene modification. Cloven hooves and cud chewing are both necessary to render an animal kosher. Genetic modifications will not render an animal that is otherwise kosher into a nonkosher animal unless it modifies those two requirements (rendering the cloven hoof into a single hoof or eliminating cud chewing)."[69]

Conclusion—Khazir: *Return to Kashrut*

As discussed above, the laws of *kilayim* are considered *khukim*—immutable statutes. "If, indeed, we are enjoined to treat the Torah's ban as a chok . . . and not to expand its parameters beyond the parameters given, then it seems that no extension to genetic techniques is warranted," concludes Reisner. However, he continues, "Most commercial considerations have ramifications which could be life-saving. . . . [p]est resistance . . . to feed the world's starving population . . . [and] greater shelf life . . . can facilitate distribution of foodstuffs to the needy. . . . [T]his would appear to be sufficient reason to permit genetic engineering to continue."[70]

Shulman strongly supports the use of genetic engineering as it eliminates some of the halakhic problems inherent in crossbreeding. "In actual fact, laboratory production of genetically modified organisms may be even more acceptable than selective breeding and crossbreeding, since, where Judaism forbids the actual process of crossbreeding, it still permits the resulting product. It is likely that, if done genetically, even the process itself would be permitted."[71]

Even though rabbinic views appear to prohibit a "kosher pig," Bleich's analysis of the babirusa provides a provocative possibility. Perhaps a kosher pig could exist after all, suggests Bleich. He cites Rabbi Hayyim ibn Attar in the *Or ha-Hayyim*. "'Why is [the pig] . . . named 'hazir' [the Hebrew word for return]? Because it will one day 'return' to become permissible.'"[72]

"The pig will return to its pre-Sinaitic status as a permitted source of meat," Bleich declares. "Accordingly, Or ha-Hayyim comments that the phrase 'but it does not chew its cud' which occurs in Leviticus . . . is conditional in nature, i.e., the pig is forbidden only so long as it does not chew its cud, 'but in the eschatological era it will chew its cud and will 'return' to become permissible.'"[73]

And, ironically, DNA technology may one day be used to produce such a pig. Then the pig may "return" to its status as permitted food. "Indeed, the etymological analysis presented by Or ha-Hayyim would lead to acceptance of a cud-chewing pig not only as a kosher animal but as a harbinger of the eschatological era as well," Bleich ponders. Rabbi Bleich offers his view that one day a kosher pig will usher in the return to Zion, and the Messianic age.[74]

And then imagine the culinary possibilities of kosher bacon, ham, and pork. Then again, we may not want to imagine that. The idea of eating pig, even kosher pig, may not sit well with the kosher consumer, even in the Messianic era. Kosher pork may go the way of *gevina*-burgers. On the other hand, perhaps Moo Shoo kosher pork will become the latest rage at kosher weddings and bar mitzvahs and take its place alongside the kosher sushi bars and kosher caviar.

CHAPTER TWELVE

Treife Tomatoes:
Brave New Plants

T HERE ARE STRANGE things going on in laboratories all over the world. Scientists are taking genes from a variety of species, including animals and microorganisms, and transferring them into plants, in the process creating new varieties with unusual traits. One of the first experiments of this kind involved taking a gene from a firefly and putting it into a tobacco plant, creating a plant that glows in the dark. This was not some mad scientist's farfetched plan for making a cigarette that lights itself. It was designed to use the firefly gene as an indicator, to prove that it was possible to make a gene from one species active in a completely different species. In fact, firefly genes and genes from other fluorescent species (e.g., jellyfish) have been used as markers and have been introduced into plants, mice, monkeys, and other organisms to serve as markers. Those types of markers help researchers to spot the animals and their structures that have incorporated and are expressing transplanted genes. When a transgenic plant or animal glows in the dark, it proves that the firefly's (or jellyfish's) genes have been activated in that organism.[1]

Genetic engineering has progressed from these initial studies using indicator genes to procedures involving genes to improve crops. Genetically modified (GM) produce includes frost-resistant fruits and vegetables made by transferring genes from frost-resistant animals, such as arctic fish.[2] The blood of arctic fish does not freeze in water that is colder than zero degrees Centigrade (the freezing point for pure water), because "antifreeze" proteins reduce the freezing point of its blood. (Ocean water remains liquid below zero Centigrade as well, since the dissolved salts and minerals in the ocean reduce its freezing point.) The presence of antifreeze proteins in transgenic plants can reduce the risk of crop damage for many species. Florida and other highly agricultural regions have experienced unusual weather patterns during the winter, threatening vulnerable species. When frost-resistant species are grown, farmers will reduce their losses resulting from crop damage by ice crystals.

There are also plants that make their own pesticides (insects die when they ingest these plants) and plants that have been altered to be herbicide resistant.[3] The latter can be grown in fields where herbicides can be applied to kill weeds, while the crop plant itself remains unaffected. Some plants have been genetically modified to resist disease organisms such as viruses. Genetically engineered plants, including corn, potatoes, rice, and a myriad of other vegetables, end up in our produce aisles and in processed foods on our grocery shelves.

The so-called "Flavr-Savr" tomato, which can ripen longer on the vine and still be shipped without rotting, was the first transgenic plant to be approved by the Food and Drug Administration (FDA). Other novel plants produced by genetic engineering include virus-resistant squash, herbicide-resistant soybean plants, potatoes that absorb less oil when fried, and insect-resistant cotton and corn.

The methods used to produce GM plants typically involve first identifying a trait that would be useful in a plant. An organism exhibiting such a trait can be used as a source of DNA. DNA is isolated from that organism (a plant, fungus, animal, or microorganism). Then, a minuscule quantity (picograms) of DNA is incorporated into a living cell of the recipient plant. This procedure can be done in several different ways. One method involves the use of a device called a gene gun. The DNA from the donor species is mixed with microscopic pellets made of gold or tungsten, and the DNA adheres to these tiny metal spheres. The gene gun propels DNA-coated particles through the plant cell walls—the rigid structures that protect plant cells and give the cell its characteristic boxy shape—and some of the DNA gets into the nuclei, becoming part of plant cell chromosomes. An alternative method involves infection of the plant cells with agrobacterium—a species of bacterium that infects plant cells by transferring its DNA to the plant. When this method is used, the normal genes of the agrobacterium can be replaced with whatever gene one wishes to transfer.[4]

DNA-treated plant cells are placed into petri dishes and cultivated in tissue culture, like microscopic seeds. Those individual cells that have taken up new DNA sequences (and are transgenic) can be selected. The transgenic cells that grow into tiny clusters of cells are next transplanted to an agar (jelly-like) growth medium in a petri dish. In that dish, plant hormones are added to stimulate the growth of roots, shoots, and leaves. The tiny plantlets are eventually transplanted to pots with soil, where they grow into full plants to be later moved into fields and grown as farm crops.

Although there are many technical difficulties in these procedures, and the monetary investment is initially high, genetic engineering to produce transgenic plants has become quite important commercially. FDA regulations do not require the labeling of foods that have been genetically altered, so we are already consuming these products unknowingly.

According to a 1999 *Consumer Reports Magazine* study, in which researchers from the consumer advocacy magazine went shopping and submitted a variety of foods for DNA analysis, "Genetically engineered ingredients were present in everything from infant formula to corn-muffin mix to McDonald's McVeggie Burgers." Products such as baby formula (Enfamil ProSorbee Soy Formula, Similac Isomil Soy Formula and Nestle Carnation Alsoy), soy burgers (e.g., Morningstar Farms Better'n Burgers), and a number of other processed foods (e.g., Bac-Os Bacon Flavor Bits, Jiffy Corn Muffin Mix, Old El Paso Taco Shells) were identified as having genetically modified ingredients. "We're not suggesting that eating these food isn't safe," indicated the article. "The results do show that genetically engineered foods are common."[5]

They are incredibly common in foods found on American grocery shelves. A recent survey reported by The *New York Times* revealed that "About 60 to 70 percent of the food in supermarkets may contain genetically engineered ingredients, according to the Grocery Manufacturers of America. These foods include cereals, soft drinks, ice cream, chocolate, and a host of other products. The reason is that about half the soybeans and a quarter of the corn grown in the United States contain genes from bacteria that make the plants resistant to herbicides or insects."[6] In addition, some vegetables and fruits, including squash and papaya, may be genetically engineered. (In Hawaii the papaya crop was devastated by a viral infection. Genetic engineering produced a virus-resistant variety, and this saved the papaya industry on those islands.[7])

Is It Permissible to Create GM Plants?

Since we are already eating products of genetic engineering, an important question for the Jewish consumer is: How do these new genetically engineered groceries affect the Jewish world? Where should we draw the line in the production of genetically modified foods? Should genetic engineering be used to improve crop yield, and ease of transport, flavor, and nutrition of plants? Should we be wary about risks to health and environment, some known, but some still unknown and impossible to anticipate?

The discussion in chapter 11 regarding genetically engineered animals addresses many of the issues involved in determining whether it is permitted by Jewish law to genetically engineer foods. In addition to those issues, there are some more important questions vis-à-vis plant genetic engineering.

The laws of *kilayim* stipulate prohibitions regarding the mixing of species, including plants and animals. Among the prohibitions are some forms of *kilayim* that apply only to the development and propagation of plant species; e.g., *kilayim* can include planting a mixture of seeds of different species together in the same field, the sowing of different species of seeds in a vineyard, or the grafting of a branch of one species onto another species. These types of agricultural procedures have been around for thousands of years, since the early days of plant domestication.

The laws of *kilayim* described in the Mishnah are incredibly specific: "Pears and small wild pears . . . are not *forbidden junction* one with the other. . . . [t]he apple and the crabapple . . . though they are similar to one another, are *forbidden junction* one with the other."[8] For plant as well as animal prohibitions, the Mishnah gives specific and very precise instructions. The acts listed as being forbidden are very detailed. Dozens of species are specified as to whether or not they may be commingled or crossbred. Just as the mixing of seeds is forbidden, the grafting of one species onto another is also forbidden. Mishnah *Kilayim* states, "One must not graft a tree onto a tree [of another species], a herb onto a herb [of a different species], or a tree onto a herb, or a herb onto a tree. Rabbi Judah permits a herb onto a tree."[9] Of course any procedure approximating DNA technology is not mentioned in the text.

All the laws in Mishnah *Kilayim* are considered to be *khukim,* or statutes for which God provides no explanation. According to tradition, *khukim* are derived directly from the Bible and must be adhered to without modification. And since these *khukim* are so specific, and no explanation is provided for their enactment, rabbis insist that each *khok* be followed precisely as described, to the letter of the law. They may neither be added to nor subtracted from. Thus, although the Mishnah directs that grafting of plants is prohibited, this does *not* necessarily prohibit the isolation of specific DNA sequences and the chemical joining of those DNA strands to DNA from another species.

With regard to prohibitions of sowing mixed seeds, it should be noted that even the carefully controlled modern seed industry does not produce bags of seeds that are entirely of one species. When large bags of seed are purchased from seed companies, the packages typically contain mixtures of seeds, including contaminating seeds that are not desired.[10] For instance, Scott's Pure Sun Premium Grass Seed (which contains 1,000,000 seeds per pound) has ingredients that include: "Other crop seed 0.74 %, Inert Matter 1.91 %, Weed seed 0.01%." Other mixtures sold list contaminating seeds as prevalent as "1.2 % other crop seed," and "0.06 % weed seed." Before seeds were commercially available, they were obtained by the farmer from the previous year's harvest. So barley seeds could very well have some wheat mixed in, or vice versa. And it would be a monumental task for a farmer to remove contaminating seed from large containers with many liters of mixed seed. Thus, although the Mishnah indicates that seeds of certain species should not be sown together, it permits the planting of a mixture of seeds if the contaminating seed (the minority in the mixture) is present in less than one-twenty-fourth of the volume of the majority seed. The Mishnah refers to the permitted ratio as less than one *rova* in one *seah*. (Note that one *seah* = 24 *rova*. A *rova* is slightly more than half a liter and a *seah* is about 13 liters.) Hence, if the mixture contains less than one part of the minority seed in twenty-four parts of the majority, the mixture may be sown.[11]

Genetic engineering involves placing minuscule (from one trillionth to one billionth of a gram) quantities of DNA into samples of plant cells and tissue weighing millions or billions of times greater. Thus, the ratio of DNA to plant cells is many orders of magnitude lower than 1/24. Since such small quantities of DNA are added to plant cells there should be no prohibition of *kilayim*. (In fact, since the DNA is not even visible to the naked eye, according to many halakhic criteria it does not exist; see chapter 11.)

It thus appears that, from a quantitative perspective, the prohibition of *kilayim* should not apply to DNA technology at all. Rabbi Avram Reisner states, "The *kashrut* laws of prohibited admixtures do not apply to the submicroscopic manipulation of genetic material. . . . 'Natural' *kilayim* products, though the fruit of an illicit operation of *kilayim*, have nonetheless been permitted as early as the Tosefta. . . . [O]f genetically engineered foodstuffs it should be minimally said that even if genetic engineering is to be prohibited, the products thereof are permissible."[12]

Is V'rappo Yirrape *a Sanction for Genetic Engineering?*

The most compelling reasons to allow genetic engineering of plants have to do with applications that will benefit medical science, feed the world's hungry, or improve the nutritional value of food. All of these applications could fall under the rubric of *V'rappo yirrape* (and heal, he shall heal). "The influential companies creating the new foods . . . claim genetic modification . . . will soon produce heartier strains that will feed the world's hungry more cheaply; cleaner, bug-resistant plants that won't need toxic spraying; 'smart' food with more nutrients and less fat, and even vegetables that will deliver vaccines."[13]

Golden rice is a case in point. It was one of the first crops developed by a company with the intention of donating the technology to third world countries where children who eat little more than rice suffer from malnutrition. In some areas of the world, white rice is a staple, and for many, it is not supplemented by other vitamin-rich foods. Children in some of these regions suffer from a higher incidence of blindness, since their diets are poor in beta-carotene. A newly developed strain, called golden rice, has been genetically engineered using genes from daffodils that produce large amounts of beta-carotene. The vitamin is "built into" the grain. (The genetic alterations also make the rice look yellow, hence the term "golden rice.") This enables children to counter their vitamin deficiencies while consuming their dietary staple.[14] (These companies could probably distribute the vitamin itself to children at a much lower cost and effort, but a program such as that would not help the industry in its promotion of the philanthropic and health benefits of genetically modified foods. It is tempting to speculate that they are aggressively promoting this product for "humanitarian reasons," because it will prove to the world the value of the new technology.)

Kashrut *Status of GM Plants*

The laws of kashrut include strict prohibitions regarding the mixing of meat and dairy products. Plants, in fact all things that grow in the earth (including mushrooms, which by scientific definition are fungi and are biologically quite distinct from plants), are considered to be kosher, and considered to be neither dairy nor meat. The vegetarian status of plants and fungi makes foods from these sources *pareve*— which means that it is permissible for observant Jews to eat them together with either dairy or meat products. Would plants genetically engineered with DNA from nonkosher animals become *treife*, or nonkosher? If DNA from kosher animals were used, would the plants lose their neutral *pareve* status?

The Monsanto Corporation developed a unique potato, called New Leaf Superior, which carries a natural bacterial pesticide normally used in organic farming.[15] This is intended to combat the Colorado potato beetle, while at the same time reducing the need for spraying of chemical pesticides. Spraying less chemical insecticides will, in turn, reduce the exposure of farm hands to these harmful chemicals and reduce the ecological damage done by spraying. "What happens if you want to create a potato that's a little bit bigger, and you put a pig gene in the potato?" asks Conservative Rabbi Joshua Finkelstein. "What if they're taking something from a nonkosher animal and putting it into a kosher fruit? Does that make the fruit nonkosher? . . . Can we eat it?" In this case, Finkelstein cited Rabbi Yechiel Epstein, author of *Arokh HaShulhan*.

Over 100 years ago, Rabbi Yechiel Michael Epstein points out that . . . if *anything* that goes in our mouth could be unkosher, then we wouldn't be able to drink water. You couldn't go to a brook and take water out of a natural brook. Why? Because there are all sorts of microbes and things that might be nonkosher in that water. So [Epstein] defines eating as having not only to do with taste, but having to do with oral stimulation. It can't just be organisms that are part of the water or the air, otherwise we wouldn't be able to breathe the air or drink the water, unless it was distilled pure water. . . . Based on this . . . it's clear that according to kashrut, you're allowed to eat that potato, even if it had part of a pig gene in it.

Finkelstein reasons that if Jewish law allows us to eat the microscopic substances in water, then the DNA should not render the food unkosher.[16]

Rabbi Nisson Shulman, presenting the perspective of the British Chief Rabbinate, also reported that "since Jewish law considers the host's character as dominant, there would be no prohibition against eating plants genetically modified with animal or human genes. They would still be considered vegetarian." He explained that as long as the recipient species is kosher, then "modified and non-modified strains would be equally kosher." However, since Jews are prohibited from deliberately harming their bodies, they may eat kosher GM foods "only if they are safe to eat, and must institute adequate safeguards to prevent harmful effects of the products of genetic modification."[17]

Dr. Fred Rosner addressed the question of speciation, the generation and definition of species. If you transfer genes from one species to another, does the recipient adopt the species identity of the donor? He cites the *etrog* as an example. An *etrog* is a citrus fruit used as a ritual symbol in celebration of *Sukkot,* the Jewish holiday of Tabernacles. An authentic, kosher *etrog* must come from a tree that has not been grafted, and observant Jews worldwide pay from $20 to more than $100 for a prime specimen to use in holiday prayers. "Would the citron, used on the Tabernacles holiday for religious purposes, lose its identity if lemon genes were introduced into it? How many transplanted lemon genes are needed to consider the citron to be a lemon? Can the rabbinic concept of *bitul* be applied to this situation?"[18] *Bitul,* also known as *batul b'shishim,* refers to the nullification of a componenet if it is present as less than one-sixtieth of the total mixture.

Rosner also poses a question that pertains to observance of the sabbatical year, known as *shmita*. In Israel, every seventh year is observed as *shmita,* a Sabbath for the land. The Jewish year 5761 (2000–2001) was a *shmita* year. This practice presents problems for farmers in Israel, since observant Jews will not consume from crops of annual plants grown during *shmita*. What if it were possible to convert annual plants into perennials using the methods of genetic engineering? "The latter are not subject to some of the laws of the sabbatical year.

Thus, perennial wheat, corn, or tomatoes would be permitted in Jewish law even if grown during the sabbatical year," explained Rosner. He concludes that "these problems and issues have not yet been decisively discussed and resolved by current rabbinic authorities."[19]

Modern rabbis have delved into a number of other novel questions relating to GM foods. For instance, Rabbi Shlomo Zalman Auerbach asked what blessing one makes on genetically engineered produce whose growth properties have been altered. Observant people recite different blessings, or *brakhot,* before eating different types of foods, to give thanks and praise to God. For grape products, the blessing (which begins "Blessed art Thou Lord our God, King of the universe") ends with *boreh peri hagafen* (who created the fruit of the vine). For fruits that grow on trees, the blessing ends with *boreh peri ha'etz* (who created the fruit of the tree), and for produce that sprouts from the earth, the blessing ends with *boreh peri ha'adama* (who created the fruits of the earth). "If genetic manipulation of a tree induces it to produce vegetables such as tomatoes, the blessing to be recited when eating the 'tomatoes' is *boreh peri ha'etz,* which is the usual blessing for fruits of a tree," ruled Auerbach.[20] This ruling confirms that the host, or recipient species, is the dominant species. It is the host that is used to determine how to judge the status of a species produced by genetic engineering.

Even if rabbis rule that it is not permissible for Jews to produce new and unusual plants by moving DNA from one species to another, Jews will most likely be permitted to use those genetically engineered species. After all, with regard to *kilayim,* although it is forbidden to produce certain crossbreeds, the products of *kilayim* are permitted for use by Jews. "According to rabbinic opinion, clearly it doesn't violate *kilayim* to use this, and we can eat that funny Monsanto potato, if we want to," concluded Finkelstein.[21]

Conservative Rabbi Shammai Engelmayer grapples with some of the questions raised here and expresses concern that genetic modification really is a form of *kilayim.* "In a sense, genetic engineering is like grafting. Thus, if it is forbidden to graft a plant to a tree, is it permissible to mix the genes of a plant with the genes of a tree? And if it is not permissible to do that, how can it be permissible to introduce animal genes into plant life? If that is not permissible, might it be permissible if the purpose is to create a more reliable food supply, thereby leading to an end to death by starvation?"[22]

Engelmayer defines the principle of *batul b'shishim* as void, if present in a quantity that is less than 1 part in 60. When DNA is used to genetically modify a species, it is added in quantities much less than 1 part in 60. "Generally speaking, if something nonkosher accidentally got into something that was kosher, and it was only 1/60th or less of the food's makeup, the food remains kosher. On the other hand, if the nonkosher food was deliberately placed there, or if it substantially changes the nature of the kosher food, either through flavor or consistency, it does not matter how minute an amount exists in the food; it has been rendered unfit for the kosher consumer." Engelmayer considers genetic modification a procedure that deliberately changes the nature of food. "Thus, if a particular gene from, say, a pig was used to improve the durability of a carrot, is that carrot still permitted to be eaten? Or should implanting a gene be considered uniquely different because it has become a part of the raw food and is not merely an added ingredient in the prepared dish?"[23]

The minuscule quantity of DNA used to change the traits of GM plants is not discernable by taste or sight or oral stimulation and is not part of any "recipe." The DNA is incorporated into the living cells of the recipient, which must process the genetic instructions and go through some stages of growth before the altered plant itself produces a new variety of fruit or vegetable.

Engelmayer is concerned, however, that DNA is, in fact, an ingredient, like gelatin, or changes the substance, like rennet. Gelatin and rennet are minor ingredients, but they substantially after the quality of food—so they can confer nonkosher status on food. "Some may argue that the animal's gene is substantially different from the animal itself and is thus permissible for use," offers Engelmayer. "That is the debate over the permissibility of gelatin (a substance extracted from boiled bones, hooves, and animal tissue) and rennet (the membrane lining the stomach of a calf, used in making cheese). Yet it is hard to accept that a gene that is a defining element of an animal's makeup somehow loses its identity when removed from the animal."[24]

Gelatin and rennet are food additives and actually mix with the final product. DNA does not play that role. Rather, DNA is akin to an architect's plan for a building. It instructs the builders how to build. The "plans" may be filed away to serve as a guide to build other buildings, but they do not hold the beams together or remain as part of the structure of the building. On the other hand, gelatin is an ingredient that remains as traces in the final product, somewhat like a nail gun whose nails represent a small fraction of the mass of the house, but are still there to hold the planks together.

Rabbi Azriel Rosenfeld, who first tackled the issue of genetic engineering and halakha over twenty years ago, considered it significant that only a trace amount of DNA is necessary to alter the traits of an organism. "Rabbi Rosenfeld contends that gene surgery might be permissible in Jewish law because genes are submicroscopic particles and no process invisible to the naked eye could be halakhically forbidden," notes Rosner. "Laws of forbidden foods do not apply to microorganisms. The priest only declares ritually unclean that which his eyes can see."[25] This principle is an important one to keep in mind, not only regarding animal and plant genetic engineering but also for the question of human gene therapy (see chapter 7).

One interesting feature of DNA is that it shares many common chemical properties and components, regardless of what organism it comes from. It is the chemical similarities found in DNA of all species that make it possible to perform genetic engineering. Pieces of DNA from bacteria can be cut and spliced to DNA from humans because all DNA has the same chemical backbone—a series of alternating sugar and phosphate groups that holds the strands together. These chemical similarities in the DNA strands permit scientists to chemically link the DNA of any species to that of any other: The DNA of an elephant to the DNA of a mouse, or the DNA from an oak tree to the DNA of an earthworm. Bacterial cells can build human proteins because they are able to translate the instructions to build proteins using the same key (the "codons") as human cells. It turns out that DNA is universal in many ways, but at the same time the DNA code is unique for different organisms.

Now that the secrets of the genetic code are being revealed, chemical methods for constructing artificial genes are available as well. It would not even be necessary to isolate the DNA from a human to introduce its trait into a goat. Based on our knowledge of human genetic sequences, scientists can construct the gene artificially from individual bases (the As, Ts, Cs, and Gs). That artificial gene can be linked to the goat chromosomes. No human cells ever need to be used in this type of experiment. No donor animal is required. When DNA is produced synthetically, and no donor animal is used, there should be no question regarding *kilayim,* as there is no admixture of species. When synthetic DNA is used to alter traits of a kosher plant or animal, there should be no problem with the kashrut of the recipient, that is, the genetically modified plant or animal.

Microbiological Systems and Kashrut

For many years now the food industry has turned to microbiological production of food additives. Many of these ingredients are generated by genetically modified microorganisms. As described in chapter 11, rennet made by genetically modified microorganisms has been accepted for use in cheese production by *mashgikhim* (supervisors trained in the laws of kosher food preparation) of prominent kashrut organizations.

Until the advent of biotechnology and the growth of that industry over the last two decades, almost all ingredients for food production came from animals or from plants. Kashrut supervisors had to ascertain the source of various food additives in order to rule on the status of various processed foods. "We are witnessing today a veritable revolution, both in the scope of food technology and in the source of food ingredients," writes Dr. Judith Leff. "Everyday, common ingredients are now increasingly being produced through the completely new methods of biotechnology. . . . Thus, bacteria, fungi or yeast can be harnessed to produce a whole array of food ingredients, or modify and control traditional fermentation processes." In a way this has made keeping kosher easier, because kashrut issues pertaining to products of fermentation may be less complex than natural ingredients. According to Leff, "virtually all processed foods include at least one ingredient produced through biotechnological processes. These foods include such staples as milk, bread, wine, cheese, oil, margarine, coffee, juices, alcoholic and non-alcoholic drinks, infant formulas, and even canned fruit and vegetables."[26]

Microorganisms are not considered unkosher, as they are neither animals nor the products of animals. When microorganisms are harnessed to produce food additives, they are grown from small colonies of cells, seeded into small flasks of nutrient broth; as the cells multiply, the culture is expanded by pouring a small volume of growing organisms into sequentially larger and larger vessels. The major problem with regard to kashrut pertains to the ingredients of the growth medium, or nutrient broth, as some of the nutritional supplements used are derived from animals.

The critical issue here vis-à-vis kashrut is the contribution of the nutrient broth to the cells. "Should microorganisms be considered as distinct entities, taking nourishment from the medium and excreting enzymes . . . like a tree taking nourishment from the soil and producing fruit?" asks Leff. "In that case, the kosher status of the nutrients is of no concern."[27] If the cells are considered to be like plants, which take their nourishment from the soil and convert it into distinctly different plant products, then there is no problem. From a scientific point of view, nutrients taken up by the microorganisms are converted via a chain of complex biochemical reactions into entirely different entities—new molecules.

"Alternatively, as is usually accepted, should microorganisms be construed as having no separate existence from the medium? In that case, the composition of the growth medium, or of the starting material . . . is of primordial importance," cautions Leff. There are some instances where microorganisms convert nutrients in the medium directly into a product used in food. In that case where the product can be directly traced back to a nonkosher precursor, there may be issues with kashrut.[28]

For the most part, however, the procedure for industrial production of kosher additives using microorganisms can include the following steps described by Leff: "Bacteria are grown and stored in a small flask containing non-kosher culture medium. The content of that flask is then poured into larger flasks, which are in turn poured into large fermentors to transform glucose into citric acid." Thus, she concludes, "a small amount of non-kosher material . . . ends up as 100,000 gallons of kosher material." When enzymes such as chymosin (rennet) are produced by microorganisms, the cells use "mainly building blocks of their own manufacture. There is no apparent connection between the nutrients in the growth medium and any characteristics of the enzyme being produced," Leff explained.[29]

On the other hand, in another case of industrial production, bacteria use the milk sugar lactose directly to derive an alcohol product. The alcohol product is considered dairy, because of the direct link between reactants and products. However, when enzymes are produced, there is no direct pathway between the lactose and the enzyme. Lactose is only serving as an energy source for the cells. Enzymes are not made from components of the lactose; they are made from amino acid building blocks. That is why enzymes made in this way, as long as they are purified to be free of lactose contamination, are considered *pareve*.

Food chemistry has become so complex in light of biotechnological advances that, in order to make sensible rulings on these types of products, kashrut authorities need to understand the biological and biochemical processes involved. Leff, who has expertise in microbiology, biotechnology, and flavor chemistry, indicated that there are few individuals with expertise in those scientific areas working in kashrut supervision. Thus, the process of kosher food supervision may be lacking a critical scientific perspective. Leff indicates that "some organizations have started a formal training program for *Mashgichim,* which includes some basic scientific knowledge and an introduction to the complexities of the halachic aspects of modern industrial *kashrus.*" However, she emphasizes that "an understanding of the scientific underpinnings of ingredient production is crucial." Hence, Leff recommends that "*kashrus* organizations . . . sponsor a rabbinical student in obtaining a degree in Food Technology."[30]

Natural Foods

Community pressure by health-food advocates in the United States has resulted in a trend toward "natural foods" as opposed to laboratory synthesized ingredients. The trend may sound positive, as it implies that more healthful products are being used. However, the use of animal-derived ingredients in natural foods has, in fact, complicated issues of kosher supervision. The availability of many substances made from microorganisms is a positive one for the kosher consumer and enables kosher certification of many foods that were previously unacceptable. The "artificial" substances that are products of microorganisms are generally judged to be kosher. However, when companies go back to using natural additives, they may resort to using animal products (which are considered "natural"). And the products containing natural ingredients of animal derivation will be problematic with regard to kashrut. Even products derived from milk (rather than lab-made) will cause problems with regard to *pareve* status (the foods will be considered dairy). So the return to natural sources will lead to the need to closely scrutinize such products to determine the origin of ingredients. In actuality, ingredients derived from animal or plant sources may not necessarily be more healthful than synthetic counterparts made in the lab. And in the shift to animal-derived natural ingredients, the kosher consumer will lose access to some products.

What Environmental Safeguards Would Judaism Require?

High-tech advances in agriculture and the development of new forms of plants and animals have led to serious concerns regarding their impact on the environment. One Cornell University research study caught the public's attention when it suggested that innocent monarch butterflies could be victims of insecticide-producing GM plants. When monarch butterfly larvae were fed milkweed leaves coated with pollen from GM plants, many appeared to succumb to a chemical genetically engineered into the pollen.[31] The risk that such pollen would kill harmless insects in the wild should be significantly lower than reported in that laboratory experiment, since butterfly larvae would never encounter such high concentrations of pollen in the wild. However, that study has fed the fires of outrage of environmental advocates who insist on careful monitoring of any possible adverse effects posed by GM plants.

Environmental concerns also abound regarding genetically modified animals. Genetically engineered salmon (see chapter 11) are a case in point. If GM fish have reproductive advantages, they could outcompete the natural species in the wild; this could change the composition of fish populations (and possibly drive the wild-type fish into extinction). And since some fish, including salmon, might escape from their breeding pools into natural bodies of water, they could travel large distances, affecting the ecological balance far and wide. This scenario could have a global impact on salmon.

Jewish sources encourage respect for the environment and emphasize the need for caution to ensure the safety of human beings and their world. In Deuteronomy 22:8, the Torah commands that, when one builds a house, he is required to erect a fence around his roof, lest someone fall from it. That passage is taken as an admonition to be cautious in all undertakings. Immediately following that passage are the prohibitions regarding three forms of *kilayim:* planting seeds of other species in a vineyard, plowing with two different species of animals yoked together, and wearing linen and wool together. (Deut. 22: 9–11). It is interesting that the warning about safety immediately precedes the laws of *kilayim,* as this suggests that the Torah views the mixing of species as potentially hazardous. Note that the building of a new house with a roof is *not* prohibited. If a "new house" symbolizes technology, then the passage "he is required to erect a fence around his roof" could be interpreted as requiring Jews to insure a reasonable level of safety in the use of any technology. And since this passage immediately precedes the statutes on prohibited mixtures, perhaps the Torah is sending a special warning concerning the integrity of species and the potential dangers of forbidden mixtures.

Deuteronomy (20:19–20) directs Jews to refrain from harming fruit trees in the vicinity when staging a siege on an enemy's city. This admonition has been broadened to prohibit any wanton destructiveness, vandalism, or wastefulness. The concept, called *ba'al tash-khit* (do not destroy) ensures that Jews demonstrate respect for property— their own and that of others. Despite this admonition, Jewish law does not prohibit the use of automobiles, even though they contribute to air pollution. It does not prohibit nuclear energy, even though there are potential risks. Just as risks must be weighed against benefits with regard to automobiles and nuclear energy, *ba'al tash-khit* should be evaluated with regard to the risks and benefits of GM organisms.

GM plants and animals could pose potential hazards to the environment and to health. One concern is that novel organisms will create an imbalance in populations, which will affect the ecosystem. Rabbi Joshua Finkelstein voiced his concern regarding environmental impact. Referring to the insect-resistant potato, he asks, "Are somehow we creating a better beetle as we create this potato? Through evolution, we know that if we introduce something into an environment that kills beetles, eventually beetles that are resistant to this will propagate and will create a stronger beetle. What is that going to do to our ecosystem, to our environment?"[32]

Another serious concern with regard to GM foods is the possibility that new products present in the foods, e.g., insecticidal proteins or antifreeze proteins, could trigger allergies in unsuspecting consumers. There are people for whom traces of peanuts can induce fatal allergic reactions, hence the warning labels on processed foods ("may contain traces of peanuts"), which have been legislated to protect such highly allergic individuals. However, raw produce does not have to be labeled. And what you see is not necessarily what you get anymore. For instance, if tomatoes were to carry fish or nut genes, people with allergies to certain types of fish proteins or nut allergens might be at risk if those substances showed up in unexpected places.

Rabbi Immanuel Jakobovits strongly cautions that risks be minimized when experimental methodologies such as these are initiated. "Without prior safeguards, there is no justification for the experiments already undertaken in this sacred sphere, and a strict moratorium should be declared on further tests until the complex moral issues involved have been thoroughly examined, and some firm ethical guidelines are established to prevent abuses and excuses incompatible with the sanctity of life and its generation." Jakobovits' warnings were written more than twenty-five years ago, before most of the applications of genetic engineering were even imagined. However, his concerns remain as timely today as they were in 1975. He concludes, "It is indefensible to initiate controlled experiments with incalculable effects on the balance of nature and the preservation of man's incomparable spirituality without the most careful evaluation of the likely consequences beforehand."[33]

Shulman also cautions, "Although genetic engineering is not technically the same as cross-breeding of species, nevertheless the spirit of the Biblical law requires that adequate safeguards be set in place so that this technique may be used in the service of human beings rather than in upsetting, harming, and even ultimately destroying the natural world." However, as long as reasonable "fences" are erected to keep people and the environment from being harmed, production of genetically modified organisms should be permitted. As Shulman pronounced, "We see no ethical difference between genetic modification of animals and plants by selective breeding, and doing so in the laboratory with new scientific techniques, provided that there are adequate and failsafe guarantees against violation of human safety and prevention of unnecessary cruelty to animals."[34]

Although we accept the use of technology, Rabbi Shabtai Rappaport correctly points out that we are still ignorant of many aspects of the intricacies of nature. And thus it is not possible to anticipate all the ramifications of our activities vis-à-vis living organisms and the environment. "Because our understanding of the organism as an entity is lacking, manipulating genes may create unpredictable monsters that could cause great harm to humanity and to the environment," maintains Rappaport. "As to the danger of creating harmful monstrosities, man's activity does create dangers."[35]

In a discussion on "Ecology and the Judaic Tradition," Robert Gordis cites the Book of Job (40:15 to 41:26), which describes massive monstrous creatures, the Behemoth and the Leviathan. God wants Job to understand that everything has a role in the universe, even ferocious creatures. Gordis explains, "It is not merely that they are not under human control . . . they are positively repulsive and even dangerous to man. Yet they too reveal the power of the Creator in a universe which is not anthropocentric but theocentric, with purposes known only to God, which man cannot fathom." In other words, the world does not revolve around humankind; every creature is important (although in Genesis, humans are given dominion over its creatures). The obligation to preserve the natural world is reinforced by Scriptural descriptions of those hideous beasts. "Man takes his place among the other living creatures, who are likewise the handiwork of God. Therefore he has no inherent right to abuse or exploit the living creatures or the natural resources to be found in a world not of his making, nor intended for his exclusive use," Gordis asserts.[36]

"The world is both a mystery and a miracle; what man cannot understand of the mystery he can sustain because of its beauty," continues Gordis. "Man is not the goal of creation and therefore, not the master of the cosmos. . . . Since the universe was not created with man as its center, neither the Creator nor the cosmos can be judged from man's vantage-point. . . ."[37]

Humans are far from omnipotent or omniscient, so prudent use of the world's resources is called for. The reasoning behind *khukim,* or statutes such as *kilayim* may never be revealed to humankind. But the fact that God has provided boundaries like the laws of *kilayim* means that we do not enjoy unrestricted power to exploit the world. From this we may learn to use technology with restraint and hope eventually to attain a more Divine understanding of how to care for our world.

CHAPTER THIRTEEN

When Science and Scripture Collide

Today we are learning the language in which God created life. We are gaining ever more awe for the complexity, the beauty, the wonder of God's most divine and sacred gift. —PRESIDENT BILL CLINTON, UPON THE ANNOUNCEMENT OF THE COMPLETION OF THE HUMAN GENOME SEQUENCE.[1]

IN OUR MODERN world, science and Scripture encounter each other in novel situations, leading to conundrums never before imagined. In this book, the analysis of Judaic reactions to biotechnology illustrates how Jewish tradition and law can be adaptable in novel circumstances. When science and Scripture "collide," neither one needs to self-destruct. In fact, I have documented many instances where biotechnological breakthroughs are consistent with Jewish tradition and law.

For instance, Jewish sources generally accept those applications of science and technology which can improve the health and well-being of human beings. Thus, new technologies that involve the treatment of infertility, and the use of stem cells or gene therapy to cure disease are sanctioned by many rabbinic sources. To be sure, there are boundaries imposed by religious values, thus, some applications of biotechnology are not permitted. In fact, some of the techniques discussed here appear to be unacceptable across the board. (One example would be using methods of sex selection to fulfill personal preference.)

This book documents Judaic views from the perspectives of Reform, Conservative, and Orthodox sources. I have shown that in many instances, with regard to bioethical issues, it may not be possible to predict how Orthodox, Conservative, and Reform rabbis will react or rule on an issue based on the assumption that more liberal Jews have more liberal views. Thus, with regard to bioethics, rabbinic rulings cannot be pigeonholed. Each issue stands on its own, and must be addressed individually by rabbinic adjudicators. Rabbinic sources frequently hold permissive views regarding technology when health issues are involved. For example, many Orthodox rabbis are known to insist on stringent observance of *pikuakh nefesh*, the preservation of life—even when it leads to violation of the Sabbath. On the other hand, there are some examples where rabbis on the more liberal end of the spectrum have appeared less willing to accept certain new technologies. In the case of human cloning, for instance, Orthodox rabbis may be more inclined than their Reform counterparts to accept human cloning as a treatment for infertility or for other health-related applications.

DNA *in Torah and Talmud*

One of the major cornerstones of Jewish bioethics (see chapter 1) is *Ayn khadash takhat hashemesh,* or "There is nothing new under the sun" (Eccles. 1:9). In my explorations of Judaic attitudes with regard to reproduction and infertility, alteration of traits, human development, and Jewish identity, I have discovered a variety of views, rabbinic opinions and rulings, and scriptural insight on modern issues. For many of these issues, although there is no scientific precedent, there are insights to be gleaned from the Torah, the Talmud, and other writings that appeared decades, centuries, or millennia before biotechnology was developed. Genetic engineering has scientific roots in ancient techniques of crossbreeding. Perhaps the bioethical responses to genetic engineering can likewise be rooted in the ancient laws and practices vis-à-vis crossbreeding. Despite the fact that some technologies are unprecedented, for many of the issues in modern biology, "there is nothing new under the sun."

If the Talmud is interpreted metaphorically, it appears that even DNA, whose scientific role as genetic material was not known until the twentieth century, may have been alluded to in that ancient text. The Talmud states: "R. Hanina b. Papa taught: 'What is the meaning of the Scriptural 'you have winnowed my going and my lying' (Ps. 139:3)? It teaches that man is not formed from the entire drop *[tipah],* but only from its clearest part."[2] In this discussion pertaining to human reproduction, the Talmud teaches that man is determined by "the clearest part." The term *tipah,* or drop, could be referring to DNA, the genetic material that determines human traits. The "entire drop" could refer to the entire human genome, and the "clearest part" would be analogous to the portions of the genome that contain functioning genes.

The Hebrew word *tipah,* or drop, is used again in another talmudic passage: "The name of the angel who is in charge of conception is 'Night,' and he takes up a drop and places it in the presence of the Holy One, blessed be He, saying, 'Sovereign of the universe, what shall be the fate of this drop? Shall it produce a strong man or a weak man, a wise man or a fool, a rich man or a poor man?'"[3] Here the term *tipah* refers to a determinant of human potential, perhaps the genes.

Other passages using *tipah* include, "When the drop appears from that a righteous man will be fashioned."[4] Referring to the plague that wiped out all Egyptian firstborn children, Rashi asserts the power of God to identify firstborn children: "I am He who chose in Egypt between the *tipah* [drop] of the firstborn and drop of the non-firstborn."[5] In both of these cases, *tipah* could be translated as "semen." However, interpreting the word *tipah* as genetic properties or potential would make even more sense in this context.

DNA Code and Torah Code

The National Human Genome Research Institute (NHGRI) of the National Institutes of Health (NIH) supports Genome Scholar Development to help researchers establish independent programs in genomic research. In this program, DNA scholars are being sought to complete the analysis of the recently completed human genome sequence (and genomes of other organisms as well). DNA scholarship could potentially provide lifelong careers for many young scientists, as it will undoubtably take decades to identify and understand the functions of all the human genes. The genome that has thus far been sequenced is a "consensus" database, which provides examples of genes but not all forms of every gene. In actuality, for each gene there can be multiple forms or alleles. Thus, in a sense, the "complete" human genome will not be available until every person's genes are sequenced and all iterations of genes are studied. And since there are new people conceived and born every day, there will be new variations of the human genome generated, thus, the identification and study of genes will continue on. That explains why the job of the DNA scholar will be open-ended, and will constantly evolve to address whatever genetic problems or questions arise in the future.

In the course of this endeavor, DNA scientists will have to learn how the genetic instructions relate to the product. Molecular biologists already know the mechanics of how the instructions (the As, Ts, Cs, and Gs) direct cells to link together amino acids in the building of protein chains. However, biochemists have known for some time that proteins themselves are amazingly complex, and cannot be understood by only learning the order of their amino acid building blocks. Proteins are three-dimensional structures and are in many ways more complex than the instructions used to make them. Just as the science of genomics studies the intricacies of DNA sequences, the science of proteomics is addressing the even more intricate protein structures. The structure and function of many of the proteins determine development, disease, and disorder in cells and organisms, and it will take many decades to understand the basis for the construction of these complex molecules.

There are some interesting analogies between the human genome and the Scriptures. Just as, in genomic studies, scholars may devote their lives to learn the intricacies of genetic code, Torah scholars can spend their entire lives learning the Torah and its commentaries and never completely master the texts. The DNA code directs the formation of the proteins, molecules which form the physical being. The Scripture, that is Torah and Talmud, also contains a code—a moral code that directs the development of the moral and spiritual being. According to traditional Jewish beliefs, God, as Creator of the Universe, is the author of both the scriptural and DNA codes. (Reform Judaism maintains that the Torah is Divinely inspired, although not written by God.)

When Torah scrolls are prepared, the *sofer,* or scribe, copies the scroll by hand, letter by letter, onto a parchment made of animal skins. Accuracy is of paramount importance. If an error is found, it must be repaired. This can sometimes be accomplished by scratching off the ink and rewriting the letter or letters. If the error is more serious, a whole section of the scroll may have to be cut out, and the corrected section sewn back into its place. This system of writing Torahs and repairing Torah scrolls, which has been adhered to for thousands of years, has kept the number of errors in the Torahs to a minimum.

The copying of DNA by cells is amazingly accurate. However, on occasion, there is an error and the wrong base is inserted. Cells have biochemical systems (repair enzymes) to detect errors and edit them out. Many mutations are therefore detected before they are passed on to the next generation. And this process has kept the number of errors in genes to a minimum. That is why mutations in the DNA are as rare as they are.

In Kabbalah, or Jewish mysticism, every letter of the Torah is imbued with meaning and value. *Kabbalah* involves the mystical interpretation of every word and letter of Torah. One system of interpretation found in Kabbalah and in the later Midrashim, called *gematriah,* is an approach to biblical interpretation based on calculating the numerical values of the Hebrew letters. For instance, a popular application of *gematriah* today is the interpretation of the Hebrew word *Chai* as its numerical value, 18 (*chet* = 8 and *yod* = 10). Based on this interpretation, when Jews give gifts or charity, they frequently give amounts in multiples of 18 (for instance, *chai,* or 18; twice *chai,* or 36; or ten times *chai,* 180). Based on principles of *gematriah,* Some modern scholars interested in mystical analysis of Scripture have used computer programs to look for hidden meaning in the ancient texts. One such analysis has tried to discern codes that predict the future.[6]

The human genome has more than 3 billion bases, or genetic letters. The coding regions of the genome contain genes that instruct how a cell must be built. Noncoding regions (more than 98 percent of the DNA) have other functions. Studies on those portions of the genome have provided insight into evolution, as the noncoding regions have retained traces of human genetic history. These areas of the genome give insight into the relationship between humans and other organisms and the process by which the human genome came to be. The so-called "junk DNA" in noncoding regions has retained remnants of ancient genes, repetitive elements and other inscrutable segments whose functions remain a mystery. DNA scholars are using computer programs to look for active genes, hidden remnants of ancient genes, and other patterns that will reveal how the genome evolved over billions of years. The genes do not predict the future, but they can give scientists a pretty good idea of the past.

DNA Codes Used to Decipher Scriptural Codes

One interesting convergence of DNA science and Judaism has occurred in a novel application of genetic technology. It involves the use of DNA technology to decipher ancient texts. Parchments made of animal skin contain DNA from the original animal skins. Many of the ancient Judaic documents, such as Torah scrolls and other biblical texts, tend to be quite long, and each scroll may be made from numerous animal skins laced together with threads made of sinew. Each segment of a scroll would have originated from a separate animal. Each skin from an individual animal has its own distinctive unique DNA, just as each goat or lamb did. When ancient documents are unearthed, they are frequently found in fragments. The fragments recovered can be tentatively matched up based on what is written on each piece, using our knowledge of modern versions of these texts or other ancient documents. However, it is often difficult or impossible to reconstruct some pieces if they are small and numerous. The task could be compared to putting together a jigsaw puzzle—from *dozens* of original puzzles whose pieces have all been mixed together. DNA analysis can help archeologists and scholars reconstruct the original manuscript by helping them sort the puzzle pieces to learn which fragments originally came from the same skin. (This is similar to organizing the puzzle pieces back into their original boxes.)

The leftover fragments from the reconstructed Dead Sea Scrolls consist of a pile of about ten thousand thumbnail-sized pieces. DNA methodology is being used to reconstruct these pieces into the eight hundred or so original scrolls they came from. "It requires investigating not the ancient scribes who wrote the texts, but their goats," reports Philip Hilt.[7] The DNA extraction method can be done without damaging the fragments, since only a tiny sample is needed. DNA is extracted from the sample and is amplified by a process called PCR, or polymerase chain reaction, which chemically copies the DNA sequence in that sample over and over again until millions of copies of each original DNA molecule have been made. This is done to provide a large enough quantity of the original DNA to enable analysis of the DNA sequence in each piece, in order to identify which pieces are from the same parchment. DNA technology has merged with scriptural analysis in this unique application.[8]

Biotechnology and Bioethics

The topics in this book have addressed cutting edge biotechnology and the new opportunities provided by this technology. Technology—or applied science—can open doors leading to new areas of knowledge previously unavailable. It can allow us to accomplish tasks never before possible. Biotechnology can facilitate great changes in agriculture, human health, family relationships, and understanding of the universe. It can also lead to new problems, unforeseen complications or catastrophes, more powerful weapons, and moral dilemmas.

Technology is neither good nor evil; it is morally neutral. What people do with technology can have positive or negative effects. This is true whether one is discussing gunpowder or automobiles or biotechnology. For instance, dynamite, invented by Alfred Nobel, can be used in construction or destruction. (Since dynamite was used in warfare, Nobel tried to atone for his work on explosives by establishing the Nobel prizes, to reward scientists whose contributions benefit humanity.)

Nuclear energy is another example of a scientific breakthrough with both positive and negative applications. Atomic bombs have caused destruction and suffering and are capable of wiping out all life on Earth. But nuclear energy—derived from radioactive materials—will most likely continue to be an important source of power in many parts of the world. And radioactive tracers and radiation therapy are used to diagnose and treat diseases, and in biomedical research studies.

We know from the lessons of history that horrendous crimes can be committed with little or no technology. During World War II, the Nazis perverted the genetic knowledge of the day, and launched a program to propagate a master race. Their methodology to propagate the master race was decidedly nontechnological; it was based on matchmaking. The Nazi government mandated the mating and reproduction of couples who were chosen and matched according to their "Aryan" traits. At the same time the Nazis used scientific research and products of technology to institute efficient programs of mass murder, in order to wipe out the undesired "inferior" races, such as Jews and Gypsies.

Sex selection has also been practiced in some societies without the benefit of high technology. No technological sophistication is needed to practice female infanticide, that is, to selectively kill female newborns. In some societies where sons are overwhelmingly preferred over daughters, such behavior has had a dramatic effect on sex ratios. In China and India both low-and high-tech methods are being used by couples who favor male offspring. As described in chapter 8, the ratio of male to female babies in China and India is significantly skewed in favor of the male. The problem has become so severe that in some areas there is a shortage of women for young men to marry. Clearly, it is not necessary to use sophisticated technology in order to wreak havoc with the world.

Ki tov: *That It Was Good*

The key to an ethical application of technology, including biotechnology, could perhaps be derived from Genesis, in the biblical account of creation. When God completes each day's work, He assesses the products of creation. "*Vaya'ar Elokim ki tov*" (And He saw that it was good). This sentence is repeated over and over pertaining to the products of each day of creation (Gen. 1: 4, 10, 11, 18, 21, 25). On the sixth day, when God creates man, the Scripture declares, "*V'hinei tov me'od*" (and behold it was *very* good) (Gen. 1: 31). God saw a qualitative difference between humankind and His other creations. Man was "very good." And humankind was given dominion over every living thing.

When humans discover and learn the laws of science and develop new technologies, we act in many ways like creators. According to Genesis, when God created man, He did so by declaring "Let us make man in Our image." Theologians have written throughout the millennia that humankind is meant to follow the example set by God—*imitatio Dei*. What God created was *ki tov*, and humans should follow God's example. Interestingly, *ki tov* is a phrase used by Eve to describe the Tree of Knowledge of Good and Evil. Perhaps this is a sign for humankind that knowledge is accessible, but must be used responsibly and with good intentions. Therefore, whatever man creates or invents, and however he uses his knowledge of the natural world, should also be '*ki tov*'—for the good.

Of Science and Faith

Maimonides' "Thirteen Principles of Faith"[9] include the belief that the resurrection of the dead will occur in the Messianic era. As a scientist, it requires a suspension of belief in the physical laws as we understand them to subscribe to that tenet of faith. One author has offered a scientific mechanism for resurrection of the dead, suggesting that human cloning could explain the physical aspects of resurrection of the dead.[10] However, I believe that some issues should remain simply faith-based, as they are not meant to be explained on a physical level.

When my mother passed away, I did not save any of my mother's cells or DNA for future cloning. She was interred on a hillside in Jerusalem and awaits the resurrection. Many Jews choose to be buried in or near Jerusalem because it is believed that those who are buried close to the site of the Holy Temple will be among the first to be resurrected. That, of course, is a tenet of faith. Cloning is neither a prerequisite for, nor will it expedite, the resurrection.

On the other hand, from a scientific perspective, I do have physical evidence that my mother's genetic legacy lives on, as her genes were transmitted to her children and grandchildren. Her genetic legacy lives on in me and in my brothers, my daughters, and my nephews. When I glimpse her traits in her grandsons and granddaughters, I see the connection—the link borne on genes. Her personal legacy also lives on— her ideals and ideas, her qualities and attributes, products of both genetics and environment, also show up in her offspring.

A genetic link is more than just a personal bond between parent and child. Research has revealed that DNA provides the link between generations, and also connects the descendants of an ancestral population to each other. There are, indeed, "Jewish genes," that many Jews retain. However, it is tradition and religious observance, ideals and ideas, that provide universal connections between all Jews and truly define the Jewish community.

The most important lesson of the Human Genome Project may be the new data emerging on commonalities within the human species. It is startling to learn that—for all our differences from person to person and race to race—each individual human differs genetically from other human beings by a mere 0.1 percent of their genes. We also share common genetic sequences with other primates: about 98 percent of our genes are identical to those of chimpanzees. We have much in common with other mammals: about 90 percent of our genes are similar to mouse genes.[11] And humans resemble lower animals in more ways than we care to admit: we share about 50 percent genetic similarity with flies and worms.[12] Thus, DNA sequences define commonalities among all living organisms.

That is quite sobering, and puts us humans in our places, not as the center of God's universe, but as servants of God. We are given a special place in the Universe—as caretakers of Earth, permitted to use resources, but not abuse them.

The *Oath of Maimonides,* meant for physicians embarking on their careers as healers, emphasizes the importance of continued research and study and the human need to learn about the universe and its natural laws. Although attributed to Maimonides, the oath was probably written by an eighteenth-century German Jewish physician, Markus Herz.[13] The oath declares, "Grant me the strength, time and opportunity always to correct what I have acquired, always to extend its domain; for knowledge is immense and the spirit of man can extend indefinitely to enrich itself daily with new requirements."[14]

The importance of applying the scientific method to understand the world is suggested by this passage in the Oath: "Today he can discover his errors of yesterday and tomorrow he can obtain a new light on what he thinks himself sure of today. Oh, God, Thou has appointed me to watch over the life and death of Thy creatures; here am I ready for my vocation and now I turn unto my calling."[15] Scientific inquiry is sanctioned by this author. Accordingly, we are permitted, or rather encouraged, to delve into the secrets and mysteries of medical knowledge.

"Knowledge is immense," states the oath. Indeed, the knowledge within our genes is so immense, it requires sophisticated computer technology to manage all the information. And it will take DNA scholars decades to decode—and attain an understanding of—the code of human physical existence.

Upon completing the sequencing of the human genome, Dr. Francis Collins proclaimed, "We have caught the first glimpses of our instruction book, previously known only to God."[16] As we read this book of humankind, we must keep in mind that we are encouraged by Jewish tradition to discover the laws and intricacies of nature. We are permitted to develop technology based on our scientific discoveries. However, we must never forget, in the course of our explorations, to remain within the bounds of ethics and human decency.

Notes

Preface (pages ix–xvi)

1. Aldous Huxley, *Brave New World* (New York, London: Harper and Brothers, 1932).

2. Talmud *Taanith* 2a (note that Talmudic citations refer to the Babylonian Talmud, unless otherwise specified), as quoted in Julius Preuss (1911), *Biblical and Talmudic Medicine,* trans. and ed. by Fred Rosner (Northvale, N.J.: Jason Aronson, 1993), 411.

Chapter 1. Introduction: Bioethics and the Jewish Spectrum (pages 1–23)

1. Samuel Gorovitz, "Bioethics and Social Responsibility," in *Contemporary Issues in Bioethics,* ed. Tom Beauchamp and LeRoy Walters (Encino, Calif.: Dickenson Pub. Co., 1978), 52.

2. J. David Bleich, *Bioethical Dilemmas* (Hoboken, N.J.: Ktav Publishing House, 1998), xii.

3. Aaron Ridley, *Beginning Bioethics* (New York: St. Martin's Press, 1998), 6.

4. Ibid.

5. Ibid.

6. Thomas A. Mappes and David DeGrazia, *Biomedical Ethics* (New York: McGraw-Hill, 1996), 44.

7. Elliot N. Dorff, *Matters of Life and Death* (Philadelphia and Jerusalem: Jewish Publication Society, 1998), 111.

8. Ibid., 110.

9. Michael A. Grodin, "Halakhic Dilemmas in Modern Medicine," *Journal of Clinical Ethics* 6.3 (1995): 218–21.

10. Mishnah *Pirkei Avot* 1:1.

11. A. Cohen, *Everyman's Talmud* (New York: E. P. Dutton & Co., 1949), xxii.

12. Ibid., xxxi.

13. Zechariah Fendel, *Anvil of Sinai* (New York: Hashkafah Pubs., 1977), 365.

14. Maimonides, *Mishnah Torah,* ed. and trans. Philip Birnbaum (New York: Hebrew Publishing Co., 1944).

15. Fendel, *Anvil of Sinai,* 376.

16. David M. Feldman, *Birth Control in Jewish Law* (New York: New York University Press, 1968), 15.

17. Grodin, "Halakhic Dilemmas in Modern Medicine."

18. Ibid.

19. Louis Flancbaum. "Using Jewish Medical Ethics to Appreciate the Relative Among the Absolute," *B'Or Ha'Torah* 12E (2001): 95–104.

20. Ibid.

21. Grodin, "Halakhic Dilemmas in Modern Medicine."

22. Dorff, *Matters of Life and Death*, 10.

23. Grodin, "Halakhic Dilemmas in Modern Medicine."

24. Central Conference of American Rabbis (CCAR), Reform Responsa 5757.2: "In Vitro Fertilization and the Status of the Embryo," 1997 <http://www.ccarnet.org/resp>.

25. Grodin, "Halakhic Dilemmas in Modern Medicine."

26. Tom Beauchamp and James F. Childress, *Principles of Biomedical Ethics* (New York: Oxford University Press, 1979), 110–11.

27. Talmud *Berachot* 33b, *Niddah* 16b.

28. David Winston "Free Will," in *Contemporary Jewish Religious Thought*, ed. Arthur Cohen and Paul Mendes-Flohr (New York: Charles Scribner's Sons, 1987), 271.

29. Mishnah *Avot* 3:15.

30. Talmud *Sanhedrin* 37a.

31. Lev. 19:16; Talmud *Sanhedrin* 73a.

32. Talmud *Baba Kamma* 85a.

33. Haim H. Cohn, "Justice," in *Contemporary Jewish Religious Thought*, ed. Cohen and Mendes-Flohr, 517.

34. Dorff, *Matters of Life and Death*, 399–400.

35. Grodin, "Halakhic Dilemmas in Modern Medicine."

36. Walter Jacobs, Central Conference of American Rabbis (CCAR), Responsa 154: "Jewish Involvement in Genetic Engineering," 1992, in *Program Guide X: Cloning*, ed. Union of American Hebrew Congregations Committee on Bio-Ethics (Philadelphia: UAHC, Summer 1998).

37. Talmud *Sabbath* 53a, *Baba Metzia* 32b, *Sanhedrin* 56a, 56b.

38. Jacobs, "Jewish Involvement in Genetic Engineering."

39. J. David Bleich, "Animal Experimentation," in *Contemporary Halachic Problems* (New York: Ktav Publishing House, 1989), 3:194–236.

40. Central Conference of American Rabbis (CCAR), "A Statement of Principles for Reform Judaism," adopted at the 1999 Pittsburgh Convention, May 26, 1999 <http://www.ccarnet.org/platforms/principles.html>.

41. A. Steinberg and J. D. Loike, "Human Cloning: Scientific, Ethical and Jewish Perspectives," *ASSIA: Journal of Jewish Medical Ethics and Halacha* 3 (1998): 11–19.

42. Mishneh Torah, *Melakhim* 6:10, as cited in Lewis Jacobs, *What Does Judaism Say About . . . ?* (New York: Quadrangle/The New York Times Book Co., 1973), 121.

43. Talmud *Sanhedrin* 73a.

44. Maimonides, *Hilchot Rotzeach u'shmirat nefesh* 1:14; Shulhan Arukh *Orach Chaim* 328:2; Talmud *Yoma* 82a; Shulhan Arukh *Yoreh Deah* 253:1, as

quoted in Immanuel Jacobovits, *Jewish Medical Ethics* (New York: Bloch Publishing Co., 1959), 54.

45. Dorff, *Matters of Life and Death*, 15.

46. Jacobs, *Program Guide X: Cloning.*

47. UAHC Committee on Bio-Ethics, *Program Guide IX: Organ Donations and Transplantation* (Philadelphia: UAHC, 1997).

48. J. David Bleich, *Contemporary Halakhic Problems* (New York: Ktav Publishing House and Yeshiva University Press, 1977), 1:93.

49. Talmud *Baba Kamma* 85a.

50. Talmud *Sanhedrin* 37a.

51. Maimonides (Rambam) Mishneh Torah, *Hilkhot Nedarim* 6:8.

52. Immanuel Jacobovits, *Jewish Medical Ethics* (New York: Bloch Publishing Co., 1959), 44.

53. Ibid.

54. *Tiferet Yisrael* on *Yadayim* (4:3). Commentary written by Israel Lipschuetz (1782–1860, Germany).

55. Dorff, *Matters of Life and Death*, 408.

56. Velvl Greene, "Ethical Issues in Community Health," in *Science in the Light of Torah,* ed. Herman Branover and Ilana Coven Attia (Northvale, N.J.: Jason Aronson, 1994), 149.

57. Dorff, *Matters of Life and Death*, 412.

58. Ibid., 413.

59. Matthew Maibaum (1986), cited in Dorff, 414.

60. Dorff, *Matters of Life and Death*, 415.

61. Biotechnology Industry Organization (BIO), *BIO Ethics* (Biotechnology Industry Organization, 1998. Website: <www.bio.org>.

Chapter 2. Fruit of the Womb (pages 24–36)

1. Dr. Vincent Brandeis, "Assisted Reproduction: Clinical Overview," presented at conference on Assisted Reproductive Techniques: Halachic, Clinical, Social, Emotional Perspectives, May 18, 1997, Mark Hotel, New York.

2. Yoel Jakobovits, "Male Infertility: Halakhic Issues," in *Be Fruitful and Multiply,* ed. Richard Grazi (Jerusalem: Genesis Jerusalem Press, 1994), 55; Minnesota Men's Health Center, Inc., Woodbury, Minn., 2000–2001, <www.mmhconline.com/articles/male_infertility.html>; Center for Male Reproductive Medicine, Los Angeles, Calif. <www.malereproduction.com/07_maleinfertility.html>.

3. Richard Amelar, Lawrence Dubin, and Patrick Walsh, *Male Infertility* (Philadelphia: W. B. Saunders Co., 1977), ix.

4. Talmud *Nidda* 30a.

5. *Tzitz Eliezer* 15:45.

6. Central Conference of American Rabbis (CCAR), Reform Responsa 5757.2: "In Vitro Fertilization and the Status of the Embryo," 1997 <http://www.ccarnet.org/resp>.

7. Maimonides, *Moreh Nevuchim* 1:72.

8. CCAR, Reform Responsa 5757.2, 1997.

9. David M. Feldman, *Birth Control in Jewish Law* (New York: New York University Press, 1968), 275.

10. Elliot N. Dorff, *Matters of Life and Death* (Philadelphia and Jerusalem: Jewish Publication Society, 1998), 101.

11. Talmud *Sotah* 11b.

12. Puah Institute brochure, undated, Puah Institute, Azriel St. 19, Jerusalem, Israel.

13. Gideon Weitzman, "New Fertility Treatments and the Halachah," lecture delivered at Congregation Keter Torah, Teaneck, New Jersey, January 31, 2001.

14. Ibid.

15. Richard Grazi and Joel Wolowelsky, "New Ethical Issues," in *Be Fruitful and Multiply,* ed. Richard Grazi (Jerusalem: Genesis Jerusalem Press, 1994), 202–3.

16. Ibid., 203–4.

17. Rabbi Moshe Heinemann, "Halachic Perspectives of Assisted Reproduction," presented at conference on Assisted Reproductive Techniques: Halachic, Clinical, Social, Emotional Perspectives, May 18, 1997, New York, Mark Hotel.

18. Moshe Tendler, *Pardes Rimonim: A Manual for the Jewish Family* (Hoboken, N.J., Ktav Publishing House, 1988).

19. David M. Feldman, *Birth Control in Jewish Law,* 3rd ed. (New York: New York University Press, 1995), 334.

20. David M. Feldman, personal communication, September 2001.

21. Feldman, *Birth Control in Jewish Law,* 334.

22. Dorff, *Matters of Life and Death,* 101.

23. Ibid.

24. CCAR, Reform Responsa 5757.2, 1997.

25. J. David Bleich, *Contemporary Halakhic Problems* (New York: Ktav Publishing House and Yeshiva University Press, 1977), 1:107–8.

26. Feldman, *Birth Control in Jewish Law,* 335.

27. Bleich, *Contemporary Halakhic Problems,* 1:107–8.

28. Feldman, *Birth Control in Jewish Law,* 335.

29. *Kallah Rabbathi* 2;52b 14, as quoted in Julius Preuss (1911), *Biblical and Talmudic Medicine,* trans. and ed. Fred Rosner (Northvale, N.J.: Jason Aronson, 1993), 412.

Chapter 3. Be Fruitful and Multiply: Male Infertility (pages 37–52)

1. Mishnah *Yevamoth* 6:6.

2. Gideon Weitzman, "New Fertility Treatments and the Halachah," lecture delivered at Congregation Keter Torah, Teaneck, New Jersey, January 31, 2001. The Puah Institute is an Israeli organization that provides counseling and religious supervision of reproductive procedures.

3. Robert Jansen, *Overcoming Infertility* (New York: W. H. Freeman, 1997), 54.

4. Jerusalem Talmud *Yevamoth* 4:5c, as quoted in Julius Preuss (1911), *Biblical and Talmudic Medicine,* trans. and ed. Fred Rosner (Northvale, N.J.: Jason Aronson, 1993), 386–87.

5. Miryam Wahrman, J. Victor Reyniak, Andrea Dunaif, Debra Sperling, and Jon Gordon, "Human Egg Pathology: Oocyte Recovery and Egg Morphology as Related to Patient Diagnosis, Fertilization Rate, and Early Development," *J. Gynecological Endocrinology* 1 (1985):12–19.

6. Mishnah *Yevamoth* 6:6.

7. *Numbers Rabbah* 10:5.

8. Talmud *Yevamoth* 64b, as discussed by Preuss, 413.

9. Mishnah *Nedarim* 11:12, as discussed by Preuss, 456.

10. Mishnah *Yevamoth* 8:2.

11. Talmud *Bekhoroth* 44b, as discussed by Preuss, 110.

12. Preuss, *Biblical and Talmudic Medicine,* 112.

13. Mishnah *Mikvaoth* 8:2–4.

14. Talmud *Niddah* 43a.

15. Talmud *Shevuoth* 18a.

16. Yoel Jakobovits, "Male Infertility: Halakhic Issues in Investigation and Management," in *Jewish Law and the New Reproductive Technologies,* ed. Emanuel Feldman and Joel B. Wolowelsky (Hoboken, N.J.: Ktav Publishing House, 1997), 123.

17. Fred Rosner, *Modern Medicine and Jewish Ethics* (Hoboken, N.J.: Ktav Publishing House, and New York: Yeshiva University Press, 1986), 103.

18. Weitzman, "New Fertility Treatments."

19. P. E. Barg, M. Z. Wahrman, B. E. Talansky, J. W. Gordon, "Capacitated, Acrosome Reacted but Immotile Sperm, When Microinjected under the Mouse Zona Pellucida, Will Not Fertilize the Oocyte, *Journal of Experimental Zoology* 237 (1986): 365–74.

20. J. D. Cassidy, "A Catholic Christian Perspective on Early Human Development," in *Jewish and Catholic Bioethics,* ed. Edmund Pellegrino and Alan Faden (Washington, D.C.: Georgetown University Press, 1999), 127–38.

21. Weitzman, "New Fertility Treatments."

22. Ibid.

23. Jansen, *Overcoming Infertility,* 133.

24. Jakobovits, "Male Infertility," 127.

25. Elliot N. Dorff, *Matters of Life and Death* (Philadelphia and Jerusalem: Jewish Publication Society, 1998), 81.

26. Ibid., 80–97.

27. Fred Rosner, "Artificial Insemination in Jewish Law," in *Jewish Bioethics,* ed. Fred Rosner and J. David Bleich (Brooklyn, N.Y.: Hebrew Publishing Co., 1979), 107.

28. Weitzman, "New Fertility Treatments."

29. Moshe Feinstein, *Iggrot Moshe, Even HaEzer,* part 1, no. 71, 1959.

30. Ibid., part 4, no. 32:5, 1981.

31. Fred Rosner, *Medicine and Jewish Law* (Northvale, N.J.: Jason Aronson, 1993), 2:30.

32. Jakobovits, "Male Infertility," 134.

33. Kenneth Brander, 2001. "Artificial Insemination and Surrogate Mother-hood through the Prism of Jewish Law," *B'Or Ha'Torah* 12E (2001): 59–65.

34. Rosner, *Modern Medicine and Jewish Ethics*, 102–3.

35. Dorff, *Matters of Life and Death*, 69–70.

36. Ibid., 70.

37. Ibid., 72.

38. Richard Grazi and Joel Wolowelsky, "New Ethical Issues," in *Be Fruitful and Multiply*, ed. Richard Grazi (Jerusalem: Genesis Jerusalem Press, 1994), 200.

39. Dorff, *Matters of Life and Death*, 79.

Chapter 4. Embryonic Stem Cells: When Does Life Begin?
(pages 53-64)

1. Tzvi Flaum, Conference of Association of Orthodox Jewish Scientists, Ellenville, N.Y., August 1998.

2. J. David Bleich, *Bioethical Dilemmas* (Hoboken, N.J.: Ktav Publishing House, 1998), 210–11.

3. Moshe D. Tendler, 2001. "Cell and Organ Transplantation: The Torah Perspective," *B'Or Ha'Torah* 12E (2001): 31–40.

4. Ibid.

5. Union of American Hebrew Congregations (UAHC), "Fetal Tissue Resolution," adopted at 62nd General Assembly, San Francisco, October 1993.

6. James Thomson, Joseph Itskovitz-Eldor, Sander S. Shapiro, Michelle A. Waknitz, Jennifer J. Swiergiel, Vivienne S. Marshall, and Jeffrey M. Jones. "Embryonic Stem Cell Lines Derived from Human Blatocysts," *Science* 282.5391 (1998): 1145–47.

7. Michael Shamblott, Joyce Axelman, Shunping Wang, Elizabeth M. Bugg, John W. Littlefield, Peter J. Donovan, Paul D. Blumenthal, George R. Huggins, and John D. Gearhart, "Derivation of Pluripotent Stem Cells from Cultured Human Primordial Germ Cells," *Proc. Nat. Acad. Sci.* 95.23 (1998): 13726–31.

8. National Bioethics Advisory Commission, *Ethical Issues in Human Stem Cell Research, Volume I: Report and Recommendation of the National Bioethics Advisory Commission* (Rockville, Md.: National Bioethics Advisory Commission, 1999), 69.

9. Ibid., 52.

10. Sarah Ramsey, "UK Government Looks to Expand Research on Embryos," *The Lancet* 356.9244, November 25, 2000.

11. J. D. Cassidy, "A Catholic Christian Perspective on Early Human Development," in *Jewish and Catholic Bioethics*, ed. Edmund Pellegrino and Alan Faden (Washington, D.C.: Georgetown University Press, 1999), 129.

12. "Italian Experts Back Human Stem Cell Cloning," Reuters Medical News, December 28, 2000 <www.medscape.com/reuters/prof/2000/12/12.29/20001228ethco01.html>.

13. "European Law May Change Regarding Stem Cell Research," Reuters

Medical News, July 10, 2000 <www.medscape.com/reuters/prof/2000/07/07.10/2000710ethc002.html>.

14. "Italian Experts Back Human Stem Cell Cloning."

15. Roger Pederson at conference on "Pluripotent Stem Cells: Properties, Therapeutic Perspectives and Ethical Issues," Annecy, France, June 30, 2000.

16. "Overexcitement on Embryo Stem Cells," editorial, *The Lancet* 356.9231, 26 August 2000.

17. "European Law May Change . . ."

18. Ramsey, "UK Government Looks to Expand . . ."

19. Department of Health and Human Services, Public Health Service, National Institutes of Health, "Draft, National Institutes of Health Guidelines for Research Involving Human Pluripotent Stem Cells," December 1999 <www.nih.gov/news/stemcell/draftguidelines.htm>.

20. Ibid.

21. Ibid.

22. Sharon Begley, "Cellular Divide," *Newsweek,* July 9, 2001, p. 22.

23. Antonio Regalado, Jill Carroll, and Laura Johannes, "Stem-Cell Scrum: Do Sixty Lines Even Exist?" *The Wall Street Journal,* August 14, 2001, p. 16.

24. Ibid.

25. Gina Kolata, "Researchers Say Embryos in Labs Aren't Available," *The New York Times,* August 26, 2001, p. 1.

26. Sharon Samber, "For Jews Debate Remains in Embryonic Stage," *New Jersey Jewish Standard,* July 27, 2001, p. 16.

27. Harvey Blitz, Nathan Diament (UOJCA); Herschel Billet, Steven Dworken (RCA); Letter to George W. Bush, July 26, 2001.

28. Yitzchok A. Breitowitz, "Halakhic Approaches to the Resolution of Disputes Concerning the Disposition of Preembryos," *Tradition* 31.1 (1996): 64–91.

29. Y. Breitowitz, "The Preembryo in Halacha," <www.jlaw.com/Articles/preemb.html>.

30. Ibid.

31. Maimonides, Mishnah Torah, *Melakhim* 6:10.

32. Breitowitz, "The Preembryo in Halacha."

33. Richard V. Grazi, *Be Fruitful and Multiply: Fertility Therapy and the Jewish Tradition* (Jerusalem: Genesis Jerusalem Press, 1994), 180–81.

34. Ibid., 181–82.

35. Central Conference of American Rabbis (CCAR), Reform Responsa 5757.2: "In Vitro Fertilization and the Status of the Embryo," 1997 <http://www.ccarnet.org/resp>.

36. Nicholas Wade, "Scientists Report Two Major Advances in Stem-Cell Research," *The New York Times,* April 27, 2001.

37. Judy Siegel, "Potential Cure for Heart Disease, Diabetes," *Jerusalem Post,* August 2, 2001, p. 1.

38. Tendler, "Cell and Organ Transplantation."

39. Sheryl Gay Stolberg, "Scientists Create Scores of Embryos to Harvest Cells," *The New York Times,* July 11, 2001, p. 1.

40. Richard Address, personal communication, January 9, 2001.

Chapter 5. Bone of My Bones and Flesh of My Flesh: Human Cloning
(pages 65–86)

1. Meiri (1249–1306 c.e.) on Talmud *Sanhedrin* 67b.
2. I. Wilmut, A. E. Schneike, J. McWhir, A. J. Kind, and K. H. S. Campbell, "Viable Offspring Derived from Fetal and Adult Mammalian Cells," *Nature* 385 (1997): 810–13.
3. Ira Levin, *The Boys From Brazil* (New York: Random House, 1976).
4. David M. Rorvik, *In His Image: The Cloning of a Man* (New York: Pocket Books, 1978).
5. Margaret Talbot, "Desire to Duplicate," *The New York Times Magazine,* February 4, 2001, p. 41.
6. Avraham Steinberg, "Cloning—Jewish Medical Ethics," Lecture delivered at Congregation Shomrei Torah, Fair Lawn, N.J., January 5, 1999.
7. Ibid.
8. Wilmut et al., "Viable Offspring."
9. Michael J. Broyde, "The Cloning," *Emunah* (Spring/Summer, 1998): 14–17.
10. A. Rosenfeld, "Judaism and Gene Design," *Tradition* 13 (1972): 71–80.
11. Fred Rosner, "Recombinant DNA, Cloning and Genetic Engineering in Judaism," *N.Y. State J. Med.* 79 (1979): 1442.
12. David Wolpe, as quoted in Barbara Trainin Blank, "The Ethics of Cloning," *Hadassah Magazine* (June/July 1997): 23–24.
13. Stephen M. Modell, "Medical Breakthroughs in Infertility and the Questions They Pose for Judaism," in *Program Guide XI: Infertility and Assisted Reproduction,* ed. Union of American Hebrew Congregations Committee on Bio-Ethics (New York: UAHC, Autumn 1999), 16–25.
14. John D. Loike and Avraham Steinberg, "Human Cloning and Halakhic Perspectives," *Tradition* 32 (1998): 31–46.
15. *Tiferet Yisrael, Yadayim* 4:3. Commentary written by Israel Lipschuetz (1782–1860, Germany).
16. Steinberg, "Cloning—Jewish Medical Ethics."
17. Ibid.
18. A. Steinberg and J. D. Loike, "Human Cloning: Scientific, Ethical and Jewish Perspectives," *ASSIA: Journal of Jewish Medical Ethics and Halacha* 3 (1998): 11–19.
19. Steinberg, "Cloning—Jewish Medical Ethics."
20. Talmud *Sanhedrin* 65b.
21. Michael Broyde, "Cloning People: A Jewish Law Analysis of the Issues," *Connecticut Law Review* 30 (1998): 503–35.
22. J. David Bleich, "Cloning: Homologous Reproduction and Jewish Law," *Tradition* 32 (1998): 47–86.
23. Michael Broyde, "Cloning People and Jewish Law: A Preliminary Analysis," *Journal of Halacha and Contemporary Society* 34 (1997): 27–65.
24. Bleich, "Cloning."
25. Ibid.

26. Broyde, "Cloning People and Jewish Law."

27. Central Conference of American Rabbis (CCAR) Reform Responsa 20: "Genetic Engineering," February 1978 <www.ccarnet.org/cgi-bin/respdisp.pr?me -20&year-carr>.

28. Ibid.

29. Sandra Blakeslee, "Watching How the Brain Works as It Weighs a Moral Dilemma," *The New York Times,* September 25, 2001, p. F3.

30. J. David Bleich, "In Vitro Fertilization: Questions of Maternal Identity and Conversion," *Tradition* 25 (1991): 82–102.

31. Broyde, "Cloning People and Jewish Law."

32. Bleich, "Cloning."

33. Broyde, "Cloning People."

34. Broyde, "Cloning People and Jewish Law."

35. Broyde, "Cloning People."

36. Loike and Steinberg, "Human Cloning and Halakhic Perspectives."

37. Broyde, "Cloning People."

38. Bleich, "Cloning."

39. Broyde, "Cloning People and Jewish Law."

40. Steinberg, "Cloning—Jewish Medical Ethics."

41. Broyde, "Cloning People and Jewish Law."

42. Bleich, "Cloning."

43. Aldous Huxley, *Brave New World* (New York, London: Harper and Brothers, 1932).

44. Elliot N. Dorff, *Matters of Life and Death* (Philadelphia and Jerusalem: Jewish Publication Society, 1998), 313–14.

45. Bleich, "Cloning."

46. Dorff, *Matters of Life and Death,* 311.

47. Ibid.

48. Ibid., 316.

49. Loike and Steinberg, "Human Cloning and Halakhic Perspectives."

50. Harvey L. Gordon, "Human Cloning and the Jewish Tradition," in *Program Guide X: Cloning,* ed. Union of American Hebrew Congregations Committee on Bio-Ethics (Philadelphia: UAHC, Summer 1998).

51. Ibid.

52. Talmud *Sanhedrin* 38a.

53. Gordon, "Human Cloning and the Jewish Tradition."

54. Huxley, *Brave New World.*

55. Noam Marans, "Cloning," lecture delivered at Symposium on the Ethical Issues of Biogenetic Research, Gerard Berman Day School, Solomon Schechter of North Jersey, Oakland, N.J., February 13, 2001.

56. Gina Kolata, "Researchers Find Big Risk of Defect in Cloning Animals," *The New York Times,* March 25, 2001, p. 1.

57. John Loike, "Cloning: Genetics and Implications for Jewish Law," lecture delivered at Synagogue on the Palisades, Fort Lee, N.J., March 25, 2001.

58. Moshe Shapiro, personal communication, March 28, 2001.

59. Marans, "Cloning."

60. Bleich, "Cloning."

61. Broyde, "Cloning People and Jewish Law."

62. Broyde, "Cloning People."

Chapter 6. The Seven Deadly Diseases (pages 87–108)

1. Saint Barnabas Hospital Jewish Genetic Disease Program, 200 South Orange Avenue, Livingston, N.J.

2. Robert Desnick, "Jewish Genetic Diseases: Recent Advances in Prevention and Treatment," lecture delivered at 19th Annual Conference on Jewish Genealogy, New York, N.Y., August 8–13, 1999.

3. Ruth Kornreich and Margaret McGovern, laboratory report, Mount Sinai School of Medicine, Genetics Testing Laboratory, Dept. of Human Genetics, One Gustave Levy Place, New York, N.Y., December 14, 2000.

4. Matt Ridley, *Genome* (New York: HarperCollins, 1999), 56–64.

5. Randi E. Zinberg, "Jewish Genetic Diseases," lecture delivered at L'Dor V'Dor, From Generation to Generation, Our Families, Our Health, A Health Education Conference, YM-YWHA of New Jersey, Wayne, N.J., October 10, 1999.

6. Ibid.

7. Shari Fallet and Roberta Ebert, report from Saint Barnabus Health Care System, Jewish Genetic Disease Program, Livingston, N.J., January 17, 2001.

8. Gideon Bach, "Prevention vs. Possible Cures in Human Genetic Disorders," lecture delivered at Hadassah Mediscope Project, Temple Emeth, Teaneck, N.J., April 10, 2000.

9. Ibid.

10. Ibid.

11. Desnick, "Jewish Genetic Diseases."

12. Marion Yanovsky and Lawrence Shapiro, undated letter from National Tay-Sachs and Allied Diseases Association to rabbis of New York Metropolitan area.

13. Ibid.

14. "Gift of Life" Card, NTSADA, 1202 Lexington Ave., #288, New York, N.Y. 10028.

15. Marion Yanovsky, "Tay-Sachs Today," lecture delivered at L'Dor V'Dor, From Generation to Generation, Our Families, Our Health, A Health Education Conference," YM-YWHA of New Jersey, Wayne, N.J., October 10, 1999.

16. Ibid.

17. Kornreich and McGovern, laboratory report.

18. Ibid.

19. Brochure (undated), Jewish Genetic Diseases Screening Program, Department of Human Genetics, Mount Sinai School of Medicine, New York, N.Y.

20. Brochure (undated), "For Patients of Ashkenazi Jewish Descent," Genzyme Genetics, Five Mountain Rd., Framingham, Mass.

21. Gary Samuels, Vice President of External Communications, Quest Diagnostics, personal communication, June 15, 2000.

22. Ibid.

23. Ibid.

24. Shari Ungerleider, personal communication, June 2000.

25. Ibid.

26. Ibid.

27. Jonathan Jarashow, *The Silent Psalms of Our Son* (Nanuet, N.Y.: Philip Feldheim, 2001).

28. Ibid.

29. Orren Alperstein Gelblum, personal communication, January 1999.

30. Ibid.

31. Jennifer Hahn, 1998, press release, Canavan Foundation, 600 West 111 St., New York, N.Y. 10025, November 11, 1998; and <www.canavanfound ation.org>.

32. Kornreich and McGovern, laboratory report.

33. < www.canavanfoundation.org>.

34. Kornreich and McGovern, laboratory report.

35. Gelblum, personal communication.

36. Ibid.

37. Ibid.

38. S. A. Slaugenhaupt, A. Blumenfeld, S. P. Gill, M. Leyne, J. Mull, M. P. Cuajungco, C. B. Liebert, B. Chadwick, M. Idelson, L. Reznik, C. M. Robbins, I. Makalowska, M. J. Brownstein, D. Krappmann, C. Scheidereit, C. Maayan, F. B. Axelrod, J. F. Gusella, "Tissue-specific Expression of a Splicing Mutation in the IKBKAP Gene Causes Familial Dysautonomia," *Am. J. Hum. Genet.* 68 (2001): 598–605.

39. S. L. Anderson, R. Coli, I. W. Daly, E. A. Kichula, M. J. Rork, S A. Volpi, J. Ekstein, B. Y. Rubin, "Familial Dysautonomia Is Caused by Mutations of the IKAP Gene," *Am. J. Hum. Genet.* 68 (2001): 753–58.

40. Slaugenhaupt et al., "Tissue-specific Expression of a Splicing Mutation."

41. Ibid.

42. Dysautonomia Foundation press release, January 22, 2001, 633 Third Ave, 12th floor, New York, N.Y. 10017.

43. Brochure (undated), Jewish Genetic Diseases Screening Program, Department of Human Genetics, Mount Sinai School of Medicine, New York, N.Y.

44. Dysautonomia Foundation press release, January 22, 2001.

45. Kornreich and McGovern, laboratory report.

46. Dysautonomia Foundation press release, January 22, 2001.

47. Moshe Tendler, "Ethical Issues of Genetic Screening: The Ethical Impact of Mastering the Gene," 38th Annual Convention of the Association of Orthodox Jewish Scientists, Ellenville, N.Y., August 9, 1998.

48. J. David Bleich, *Judaism and Healing: Halakhic Perspectives* (Hoboken, N.J.: Ktav Publishing House, 1981), 106 and n.4.

49. J. David Bleich, *Contemporary Halakhic Problems* (New York: Ktav Publishing House and Yeshiva University Press, 1977), 1:113.

50. Fred Rosner, "Genetic Screening, Genetic Therapy and Cloning in Judaism," *B'Or Ha'torah* 12E (2001): 17–29.

51. Bleich, *Contemporary Halakhic Problems,* 115.

52. Rosner, "Genetic Screening."

53. Ibid.

54. Y. Zilberstein, Responsum to Richard Grazi, Shevat 5751 (February 1991), as cited in Richard V. Grazi, *Be Fruitful and Multiply: Fertility Therapy and the Jewish Tradition* (Jerusalem: Genesis Jerusalem Press, 1994), 189.

55. Grazi, *Be Fruitful and Multiply,* 189.

56. David Feldman, *Birth Control in Jewish Law* (New York: New York University Press, 1968).

57. Ibid., 253.

58. Ibid., 255.

59. Ibid., 291–92.

60. David Feldman, *Birth Control in Jewish Law,* 3rd ed. (New York: New York University Press, 1995), 323.

61. Ibid., 338.

62. Ibid., 339.

63. Ibid., 331.

64. Marrick Kukin, "Tay Sachs and the Abortion Controversy," *Journal of Religion and Health* 20.3 (1981): 224–42.

65. Morrison David Bial, *Liberal Judaism at Home* (New York: Union of American Hebrew Congregations, 1971), 13.

66. Bernard M. Zlotowitz, "Genetic Testing—Linkage Analaysis," in *Program Guide V: Genetic Screening and the Human Genome Project* ed. Union of American Hebrew Congregations Committee on Bio-Ethics (Philadelphia: UAHC, Spring 1992), 7–9.

67. Bleich, *Contemporary Halakhic Problems,* 113.

68. Bleich, *Judaism and Healing,* 106.

69. Fred Rosner, "Rabbi Moshe Feinstein," in *Pioneers in Jewish Medical Ethics,* ed. F. Rosner (Northvale, N.J.: Jason Aronson, 1997), 89–90.

70. Avraham Steinberg, "Rabbi Shlomo Zalman Auerbach," in *Pioneers in Jewish Medical Ethics,* ed. Rosner, 104.

71. Avraham Steinberg, "Rabbi Eliezer Yehudah Waldenberg," in *Pioneers in Jewish Medical Ethics,* ed. Rosner, 179.

72. Kukin, "Tay-Sachs and the Abortion Controversy."

73. Zlotowitz, "Genetic Testing—Linkage Analysis."

74. Zinberg, "Jewish Genetic Diseases."

75. Gelblum, personal communication.

Chapter 7. Designer Genes, Designer Kids *(pages 109–25)*

1. M. Wahrman, J. Reyniak, A. Dunaif, D. Sperling, and J. Gordon, "Human Egg Pathology: Oocyte Recovery and Egg Morphology as Related to Patient Diagnosis, Fertilization Rate, and Early Development," *J. Gynecological Endocrinology* 1 (1985): 12–19.

2. Neri Laufer, Hadassah Medical Center, Jerusalem, Israel, personal communication, February 1998.

3. Ibid.

4. Ibid.

5. Yitzchok A. Breitowitz, "Halakhic Approaches to the Resolution of Disputes Concerning the Disposition of Preembryos," *Tradition* 31.1 (1996): 64–91.

6. Ricki Lewis, *Human Genetics: Concepts and Applications,* 4th ed. (New York: McGraw-Hill, 2001), 208.

7. Ibid.

8. R. N. Slotnick, and J. E. Ortega, "Monoamniotic Twinning and Zona Manipulation: A Survey of U.S. IVF Centers Correlating Zona Manipulation Procedures," *J. Assist. Reprod. Genet.* 13.5 (1996): 381–85.

9. Fred Rosner, "Genetic Screening, Genetic Therapy and Cloning in Judaism," *B'Or Ha'torah* 12E (2001): 17–29.

10. Ira S. Youdovin, "The Human Genome Project (N.I.H.): A Jewish Perspective," in *Program Guide V: Genetic Screening and the Human Genome Project* ed. Union of American Hebrew Congregations Committee on Bio-Ethics (Philadelphia: UAHC, Spring 1992), 15–19.

11. Stephen Modell, "Analysis of Four Cloning Scenarios from the Perspectives of Science and the Jewish Tradition," in *Program Guide X: Cloning,* ed. Union of American Hebrew Congregations Committee on Bio-Ethics (Philadelphia: UAHC, Summer 1998), 5–9.

12. Azriel Rosenfeld, "Judaism and Gene Design," in *Jewish Bioethics,* ed. F. Rosner and J. D. Bleich (Brooklyn, N.Y.: Hebrew Publishing Co., 1979), 401–8.

13. Rosner, "Genetic Screening."

14. Rosenfeld, "Judaism and Gene Design."

15. Ibid.

16. Talmud *Baba Metzia* 84a, and *Berachoth* 20a, as cited by Rosefeld, "Judaism and Gene Design."

17. Rosner, "Genetic Screening."

18. Youdovin, "The Human Genome Project."

19. Fred Rosner, "Genetic Engineering and Judaism," in *Jewish Bioethics,* ed. F. Rosner and J. D. Bleich (Brooklyn, N.Y.: Hebrew Publishing Co., 1979), 409–20.

20. Ibid.

21. Mordechai Halperin, "Human Genome Mapping: A Jewish Perspective," *ASSIA: Journal of Jewish Medical Ethics and Halacha* 3.2 (1998): 30–33.

22. Shabtai Rappaport, "Genetic Engineering: Technology, Creation and Interference," *ASSIA: Journal of Jewish Medical Ethics and Halacha* 3.1 (1997): 3–4.

23. Ibid.

24. Ibid.

25. M. Winerip, "Fighting for Jacob," *The New York Times Magazine,* December 6, 1998.

26. Website URL: <www.canavan.org>.

27. "Statement on the Death of Jesse Gelsinger," Institute for Human Gene Therapy, University of Pennsylvania, May 25, 2000. Website URL: <health.upenn .edu/ihgt/jesse.html>.

28. M. Cavazzana-Calvo, S. Hacein-Bey, G. De Saint Basile, F. Gross, E. Yvon, P. Nusbaum, F. Selz, C. Hue, S. Certain, J.-L. Casanova, P. Bousso, F. Le-Deist, and A. Fisher, "Gene Therapy of Human Severe Combined Immunodeficiency (SCID)—XI Disease," *Science* 288 (2000): 669–72.

29. M. A. Kay, C. S. Manno, M. V. Ragni, P. J. Larson, L. B. Couto, A. McClelland, B. Glader, A. J. Chew, S. J. Tai, R. W. Herzog, V. Arruda, F. Johnson, C. Scallan, E. Skarsgard, A. W. Flake, K. A. High, "Evidence for Gene Transfer and Expression of Factor IX in Haemophilia B Patients Treated with an AAV Vector," *Nature Genetics* 24(2000): 257–61.

30. Website URL: <www.canavan.org>.

31. Judy Siegel-Itzkovich, "How a Hole in the Head Made Medical History," *The Jerusalem Post* (Internet Ed.), October 15, 2001. Website URL: <www.jpost .com/Editions/2001/10/14/Health>.

32. Youdovin, "The Human Genome Project."

Chapter 8. Chosen Children: Sex Selection (pages 126–40)

1. Talmud *Berakhot* 5b.

2. Talmud *Niddah* 31b.

3. E. F. Fugger, S. H. Black, K. Keyvanfar, J. D. Schulman, "Births of Normal Daughters after MicroSort Sperm Separation and Intrauterine Insemination, In Vitro Fertilization, or Intracytoplasmic Sperm Injection," *Human Reproduction* 13 (1998): 2367–70.

4. Talmud *Kiddushin* 82b.

5. Talmud *Baba Bathra* 10b.

6. Beth Hess, Elizabeth W. Markson, and Peter J. Stein, *Sociology,* 5th ed. (Needham Heights, Mass.: Allyn & Bacon, 1996), 150.

7. Ganapati Mudur, "Indian Medical Authorities Act on Antenatal Sex Selection," *British Medical Journal* 319 (1999): 401.

8. Meredith Wadman, "So You Want a Girl?" *Fortune,* February 19, 2001, pp. 174–82.

9. R. Ramachandran, "In India, Sex Selection Gets Easier," *The UNESCO Courier* 52 (1999): 29.

10. Mudur, "Indian Medical Authorities."

11. Dr. Malpani, Malpani Infertility Clinic, Bombay, India. Website URL: <www.howtohaveaboy.com> and <www.drmalpani.com>.

12. P. Liu, and G. A. Rose, "Social Aspects of More than 800 Couples Coming Forward for Gender Selection of Their Children," *Human Reproduction* 10.4 (1995): 968–71.

13. Sonni Effron, "Japanese Couples Think Pink," *The Los Angeles Times,* November 15, 1999, p. 1.

14. For example, Talmud *Niddah* 31a.

15. Talmud *Berachoth* 60a.

16. *Da'at Zekenim mibaalei haTosafot,* on Lev. 12:2.

17. I Chronicles 8:40.

18. Talmud *Niddah* 31a.

19. Talmud *Niddah* 31b.

20. Talmud *Berakhot* 60a.

21. Julius Preuss (1911), *Biblical and Talmudic Medicine,* trans. and ed. Fred Rosner (Northvale, N.J.: Jason Aronson, 1993), 389.

22. Hippocrates, who lived in Greece from 460 to 377 B.C.E., is considered by many as the greatest physician of ancient times; Galen, a physician and philosopher, lived in Rome from 130 to 200 C.E. and produced more than 500 tractates on medicine, philosophy, and ethics.

23. Preuss, *Biblical and Talmudic Medicine,* 390.

24. Fred Rosner, *Sex Ethics in the Writings of Moses Maimonides* (Northvale, N.J.: Jason Aronson, 1994), 49.

25. Preuss, *Biblical and Talmudic Medicine,* 391.

26. Rosner, *Sex Ethics,* 76.

27. Celia W. Dugger, "Modern Asia's Anomaly: The Girls Who Don't Get Born, *The New York Times,* May 6, 2001, p. wk 4.

28. Effron, "Japanese Couples Think Pink."

29. R. Ericsson, C. Langevin, and M. Nishino, "Isolation of Fractions Rich in Human Y Sperm," *Nature* 246 (1973): 421–24.

30. J. H. Check and D. Katsoff, "A Prospective Study to Evaluate the Efficacy of Modified Swim-up Preparation for Male Sex Selection," *Human Reproduction* 8 (1993): 211–14; M. J. Chen, H. F. Guv, and E. S. Ho, "Efficiency of Sex-Selection of Spermatozoa by Albumin Separation Method Evaluated by Double-Labelled Fluorescence In-Situ Hybridization," *Human Reproduction* 12 (1997): 1920–26; C. J. DeJonge, S. P. Flaherty, A. M. Barnes, N. J. Swann, C. D. Matthews, "Failure of Multitube Sperm Swim-up for Sex Preselection," *Fertility and Sterility* 67 (1997): 1109–14; O. Samura, N. Mihara, H. He, E. Okamoto, K. Ohama, "Assessment of Sex Chromosome Ratio and Aneuploidy Rate in Motile Spermatoza Selected by Three Different Methods," *Human Reproduction* 12 (1997): 2437–42; G. A. Rose and A. Wong, "Experiences in Hong Kong with the Theory and Practice of the Albumin Column Method of Sperm Separation for Sex Selection," *Human Reproduction* 13 (1998): 146–49.

31. W. E. Richards, S. M. Dobin, V. Malone, A. B. Knight, T. J. Kuehl, "Evaluating Sex Chromosome Content of Sorted Human Sperm Samples with Use of Dual-color Fluorescence in Situ Hybridization," *American Journal of Obstetrics and Gynecology* 176 (1997): 1172–78.

32. Fugger et al., "Births of Normal Daughters."

33. Ibid.

34. Genetics and IVF Institute website, 1998–2001: <www.microsort.net>.

35. Rakha Matkin, Genetics and IVF Institute, personal communication, February 2001.

36. Fugger et al., "Births of Normal Daughters."

37. A. Shushan and J. G. Schenker, "Prenatal Sex Determination and Selection,"

Human Reproduction 8 (1993): 1545–49; E. S. Sills, D. Goldschlag, D. P. Levy, O. K. Davis, Z. Rosenwaks, "Preimplantation Genetic Diagnosis: Considerations for Use in Elective Human Embryo Sex Selection," *Journal of Assisted Reproduction and Genetics* 16 (1999): 509–11.

38. Gina Kolata, "Fertility Ethics Authority Approves Sex Selection," *The New York Times,* September 28, 2001, p. A 16.

39. Ibid.

40. Ibid.

41. P. Liu and G. A. Rose, "Sex Selection: The Right Way Forward," *Human Reproduction* 11.11 (1996): 2343–45.

42. G. Pennings, "Family Balancing as a Morally Acceptable Application of Sex Selection," *Human Reproduction* 11.11 (1996): 2339–43.

43. Genetics and IVF Institute website: <www.microsort.net/Qualify.htm>.

44. J. David Bleich, *Judaism and Healing: Halakhic Perspectives* (Hoboken, N.J.: Ktav Publishing House, 1981), 110–13.

45. Elliot N. Dorff, *Matters of Life and Death* (Philadelphia, and Jerusalem: Jewish Publication Society, 1998), 155.

46. Shmuel Goldin, quoted in Miryam Wahrman, "Too Many Choices?" *The Jewish Standard,* October 16, 1998, p. 8.

47. Richard V. Grazi, *Be Fruitful and Multiply: Fertility Therapy and the Jewish Tradition* (Jerusalem: Genesis Jerusalem Press, 1994), 186.

48. J. David Bleich, *Judaism and Healing: Halakhic Perspectives* (Hoboken, N.J.: Ktav Publishing House, 1981), 110.

49. J. David Bleich, "Sex Preselection," in *Jewish Bioethics,* ed. F. Rosner and J. D. Bleich (Hoboken, N.J.: Ktav Publishing House, 2000), 91–98.

50. David Feldman, in Wahrman, "Too many choices?"

51. Ibid.

52. Ibid.

53. Lauran Neergaard, "Ethicist Approves of Some Sex Selection," *The Bergen Record,* September 29, 2001, p. A12.

54. Name has been changed; as quoted in Wahrman, "Too Many Choices?"

55. Lee Silver, in Wahrman, "Too Many Choices?"

56. Ruth Macklin, in Wahrman, "Too Many Choices?"

57. Feldman, in Wahrman, "Too Many Choices?"

Chapter 9. TAG A CAT: Jewish Genes and Genealogy (pages 141–65)

1. David Baltimore, "Our Genome Unveiled," *Nature* 409 (2001): 814–16.

2. Francis Collins, February 10, 2001, quoted Rick Weiss, "Human Genetic Code Complex in Its Simplicity," *The Bergen Record,* February 11, 2001.

3. Rebecca L. Cann, "In Search of Eve," *The Sciences* (September/October 1987): 30–37; T. M. Powledge and M. Rose, "The Great DNA Hunt," *Archeology* (September/October 1996): 36–44.

4. Mishnah *Yevamoth* 11:5.

5. Yaakov Kleiman, "The DNA Chain of Tradition," *Jewish Action* (Winter 5760/1999).

6. Karl Skorecki, Sara Selig, Shraga Blazer, Robert Bradman, Neil Bradman, P. J. Waburton, Monica Imajlowicz, and Michael Hammer, "Y Chromosomes of Jewish Priests," *Nature* 385 (1997): 32.

7. Mark G. Thomas, Karl Skorecki, Haim Ben-Ami, Tudor Parfitt, Neil Bradman, and David Goldstein, "Origins of Old Testament Priests," *Nature* 394 (1998): 138–40.

8. Ibid.

9. Ibid.

10. Tudor Parfitt, *Journey to the Vanished City* (New York: St. Martin's Press, 1992).

11. Mark G. Thomas, Tudor Parfitt, Deborah A. Weiss, Karl Skorecki, James F. Wilson, Magdel le Roux, Neil Bradman, and David B. Goldstein, "Y Chromosomes Traveling South: The Cohen Modal Haplotype and the Origins of the Lemba—the 'Black Jews of Southern Africa,'" *American Journal of Human Genetics* 66 (2000): 674–86.

12. Janine Lazarus, "At the Jewish Doorstep in Africa," *Hadassah Magazine* (January 2001).

13. Nicholas Wade, "DNA Backs a Tribe's Tradition of Early Descent from the Jews," *The New York Times,* May 9, 1999, p. 1.

14. Thomas et al., "Y Chromosomes Traveling South."

15. Jared Diamond, "Who Are the Jews?" *Natural History* (November 1993): 12–19.

16. Ibid.

17. Ibid.

18. Ibid.

19. M. F. Hammer, A. J. Redd, E. T. Wood, M. R. Bonner, H. Jarjanazi, T. Karafet, S. Santachiara-Benerecetti, A. Oppenheim, M. A. Jobling, T. Jenkins, H. Ostrer, and B. Bonne-Tamir, "Jewish and Middle Eastern Non-Jewish Populations Share a Common Pool of Y-Chromosome Biallelic Haplotypes," *Proceedings of the National Academy of Science* 97 (2000): 6769–74.

20. Ibid.

21. Ibid.

22. Ibid.

23. "Beta Israel," The New Standard Jewish Encyclopedia, ed., Geoffrey Wigoder (New York: Facts on File, 1992), 146–47.

24. G. Lucotte and P. Smets, "Origins of Falasha Jews Studied by Haplotypes of the Y Chromosome," *Human Biology* 71 (1999): 989–93.

25. James R. Ross, *Fragile Branches: Travels Through the Jewish Diaspora* (New York: Riverhead Books, 2000), 8.

26. Hammer et al., "Jewish and Middle Eastern Non-Jewish Populations."

27. U. Ritte, E. Neufeld, M. Broit, D. Shavit, and U. Motro, "The Differences Among Jewish Communities—Maternal and Paternal Contributions," *Journal of Molecular Evolution* 37 (1993): 435–40.

28. Lucotte and Smets, "Origins of Falasha Jews."

29. Thomas et al., "Y Chromosomes Traveling South."

30. Wendy Elliman, "Menashe's Children Come Home," *Hadassah Magazine* (October 1999); Gabe Levinson, "Remote Destinations," *The Jewish Week,* October 27, 2000, p. 43.

31. Ross, *Fragile Branches,* 152.

32. Ibid., 149–79.

33. Lazarus, "At the Jewish Doorstep in Africa."

34. Talmud *Niddah* 31a.

35. Dov I. Frimer, "Establishing Paternity by Means of Blood Type Testing," *ASSIA: Journal of Jewish Medical Ethics and Halacha* 1.2 (1989): 20–35.

36. Auerbach, as quoted in ibid., 23.

37. Frimer, "Establishing Paternity."

38. Ibid.

39. Rabbi Richard Address, personal communication, March 21, 2001.

40. Mishnah *Yevamoth* 16:3.

41. Mishnah *Yevamoth* 16:5–6.

42. Jay Levinson, "An Halakhic Reconsideration of Victim Identification," *ASSIA: Journal of Jewish Medical Ethics and Halacha* 4.1 (February 2001): 55–57.

43. Ibid.

44. Moshe Lazarus, Reuven Subar, Avrohom Lefkowitz, 1994. "Ask the Rabbi," Ohr Somayach Institutions, Jerusalem, Israel, November 12, 1994, no. 42. Website URL: <http://ohr.org.il/ask/ask042.htm>.

45. Jay Levinson, "Halachic Authority Limits Use of DNA Evidence for Identification," *The Jewish Press,* February 23, 2001, p. 10.

46. Katy McLaughlin, "Confronting Disaster Halachically," *Jewish Renaissance Media,* September 25, 2001. Website URL: <ww.jewish.com/terror/disaster 0925.shtml>.

47. Mishnah *Yevamoth* 16:6.

48. Leonard Nimoy, "So Human," *Reform Judaism Magazin* (August 1999).

Chapter 10. Judging Genes (pages 166–86)

1. Moshe Tendler, "Ethical Issues of Genetic Screening: The Ethical Impact of Mastering the Gene," 38th Annual Convention of the Association of Orthodox Jewish Scientists, Ellenville, N.Y., August 9, 1998.

2. Johns Hopkins University School of Medicine, Epidemiology-Genetics Program in Psychiatry, Winter 1999 Newsletter. Website URL: <ww.med.jhu.edu/ epigen/winter_1999.htm>.

3. Cindy Hunter, personal communication, April 19, 2001, and www.med .jhu.edu/epigen/current_studies.htm>.

4. Moshe Tendler, "Survey: What do you think?" *Moment Magazine* (April 1999): 32.

5. Johns Hopkins University School of Medicine, Epidemiology-Genetics Program in Psychiatry. Website URL: <www.med.jhu.edu/epigen/ashkenazim.htm>.

6. Ibid.

7. Tendler, "Survey."

8. Mandell Ganchrow, "Survey: What do you think? *Moment Magazine* (April 1999): 32.

9. Lois Waldman, "Survey: What do you think?" *Moment Magazine* (April 1999): 32.

10. Ricki Lewis, *Human Genetics: Concepts and Applications,* 4th ed. (New York: McGraw-Hill, 2001), 246.

11. Ibid., 247.

12. Kenneth Offit, "Founder Mutations Predisposing to Cancer in Ashkenazim: From Laboratory to the Clinic," lecture delivered at American Jewish Congress Third Leadership Conference on Jewish Women's Health Issues, "Cancer Genetics in the Ashkenazi Community: Five Years Later—What Have We Learned?" New York, September 25, 2000.

13. Lewis, *Human Genetics,* 355; Jon Palfreman, "Harvest of Fear," *Frontline/Nova,* WGBH Educational Foundation, Boston. Original airdate: April 23, 2001.

14. Richard M. Goodman, "A Perspective on Genetic Diseases Among the Jewish People," in *The 1986 Jewish Directory and Almanac,* ed. Ivan Tillem (New York: Pacific Press, 1986), 150–61.

15. Ibid.

16. Ibid.

17. Ibid.

18. Israeli Government Central Bureau of Statistics. Website URL: <www.cbs.gov.il>.

19. Tendler, "Ethical Issues of Genetic Screening."

20. J. P. Struewing, D. Abeliovich, T. Peretz, "The Carrier Frequency of the BRCA1 185delAG Mutation Is Approximately 1 Percent in Ashkenazi Jewish Individuals," *Nature Genetics* 11 (1995): 198–200.

21. C. I. Szabo and M. C. King, "Inherited Breast and Ovarian Cancer," *Human Molecular Genetics* 4 (1995): 1811–17.

22. R. Wooster, G. Gignell, J. Lancaster, S. Swift, and S. Deal, "Identification of the Breast Cancer Susceptibility Gene BRCA2," *Nature* 378 (1995): 789–99.

23. Memorial Sloan-Kettering Cancer Center (MSKCC), "More Ashenazi Jews Have Gene Defect That Raises Inherited Breast Cancer Risk," press release, New York, N.Y., October 1, 1996.

24. Judy Garber, "A 40-Year-Old Woman with a Strong Family History of Breast Cancer," *JAMA* 282 (1999): 1953–60.

25. National Institutes of Health, "Questions and Answers for Estimating Cancer Risk in Ashkenzi Jews," May 14, 1997. Website URL: <http://rex.nci.nih.gov/massmedia/backgrounders/ashkenazi.html>.

26. Sata Gopan and Kenneth Offit, "Lifetime Risks of Breast Cancer in Ashkenazi Jewish Carriers of BRCA1 and BRCA2 Mutations," *Cancer Epidemiology Biomarkers and Prevention* 10 (2001): 467–73.

27. A. A. Langston, K. E. Malone, J. D. Thompson, J. R. Daling, and E. A. Ostrander, "BRCA1 Mutations in a Population-Based Sample of Young Women with Breast Cancer," *New England Journal of Medicine* 334 (1996): 137–42.

28. Jessica G. Mandell, "New York Breast Cancer Study," brochure, Sarah Lawrence College, Bronxville, N.Y.

29. Jessica Mandell, "New York Breast Cancer Study," lecture delivered at American Jewish Congress Third Leadership Conference on Jewish Women's Health Issues, "Cancer Genetics in the Ashkenazi Community: Five Years Later—What Have We Learned?" New York, September 25, 2000.

30. HIPAA (Health Insurance Portability and Accountability Act), "Guide to HIPAA, What the Health Insurance Reform Law Means," brochure, National Partnership for Women and Families, Washington, D.C., 1998.

31. Elliot N. Dorff, *Matters of Life and Death* (Philadelphia, and Jerusalem: Jewish Publication Society, 1998), 157.

32. Ibid.

33. Dorff, *Matters of Life and Death,* 158.

34. Ibid.

35. Ibid., 160.

36. Ari Mosenkis, "Genetic Screening for Breast Cancer Susceptibility: A Torah Perspective," *Journal of Halacha and Contemporary Society* 34 (1997): 5–26.

37. Elsa Reich, "Experience of Genetic Testing: Panel Discussion," at American Jewish Congress Third Leadership Conference on Jewish Women's Health Issues, "Cancer Genetics in the Ashkenazi Community: Five Years Later—What Have We Learned?" New York, September 25, 2000.

38. Talmud *Yevamoth* 64b, cited by Immanuel Jakobovits, *Jewish Medical Ethics* (New York: Bloch Publishing Co., 1959), 155.

39. Jakobovits, *Jewish Medical Ethics,* 156.

40. J. David Bleich, *Judaism and Healing: Halakhic Perspectives* (Hoboken, N.J.: Ktav Publishing House, 1981), 106.

41. Ibid., 88.

42. Dorff, *Matters of Life and Death,* 163–64.

43. Mosenkis, "Genetic Screening for Breast Cancer Susceptibility," 8.

44. Ibid., 15.

45. *Kohelet Rabbah* 5:6; Mosenkis, "Genetic Screening for Breast Cancer Susceptibility," 16–17.

46. The Breast Cancer Linkage Consortium, "Cancer Risks in BRCA2 Mutation Carriers," *Journal of the National Cancer Institute* 91 (1999): 1310–16.

47. American College of Gastroenterology (ACOG)/Medscape Wire, "Positive Family History of Colon Cancer Should Guide Screening Decisions Among Ashkenazi Jews," October 11, 1998.

48. N. B. Freimer, in Christina Chen, "Population-wide Gene Searches," UCSF News Archive: February 1, 1997. Website URL: <www.ucsf.edu/~adcom/listserv/ucsfnews/0007.html>.

49. Siamak Baharloo, Paul A. Johnston, Susan K. Service, Jane Gitschier, and Nelson B. Freimer, "Absolute Pitch: An Approach for Identification of Genetic and Nongenetic Components," *American Journal of Human Genetics* 62 (1998): 224–31.

50. Joanna Raboy, " Survey: What Do You Think?" *Moment Magazine* (April 1999): 32.

51. L. Eppstein, "UCSF Researchers Seek Subjects for Perfect-Pitch Study," *Jewish Bulletin of Northern California,* July 25, 1997.

52. Baharloo et al., "Absolute Pitch."

53. Siamak Baharloo, Susan Service, Neil Risch, Jane Gitschier, and Nelson Freimer, "Familial Aggregation of Absolute Pitch," *American Journal of Human Genetics,* 67 (2000): 755–58.

54. Andy Evangelista and Jennifer O'Brien, "Researchers Use Web to Aid Musical Gene Hunt," *UCSF Daybreak News,* University of California, February 28, 1998. Website URL: <itssrvi.ucsf.edu/daybreak/1998/02/206_musi.htm>.

55. A *mohel* is a Jewish man specially trained to perform ritual circumcisions.

Chapter 11. Kosher Pork: Brave New Animals (pages 187–208)

1. R. D. Palmiter, R. L. Brinster, R. E. Hammer, M. E. Trumbauer, M. G. Rosenfield, N. C. Birnberg, and R. M. Evans, "Dramatic Growth of Mice That Develop from Eggs Microinjected with Metallothionein-Growth Hormone Fusion Genes," *Nature* 300 (1982): 611–15.

2. Rathin Das, "Production of Therapeutic Proteins from Transgenic Animals," *American Biotechnology Laboratory* 19.2 (February 2001): 60–64.

3. Ibid.

4. Ibid.

5. Carol K. Yoon, "If It Walks and Moos like a Cow, It's a Pharmaceutical Factory," *The New York Times,* May 1, 2000, p. A20.

6. Research Update, "Summary of a Presentation by Mark L. Brantly, M.D.," *Texas Alpha Research Day,* March 4, 2000 (Alpha-1 Foundation, Miami, Florida. Website URL: <www.alphaone.org>).

7. B. Jost, J.-L. Vilotte, I. Duluc, J.-L. Rodeau, and J.-N. Freund, "Production of Low-Lactose Milk by Ectopic Expression of Intestinal Lactase in the Mouse Mammary Gland," *Nature Biotechnology* 17 (1999): 160–64.

8. Bruce Whitelaw, "Toward Designer Milk, *Nature Biotechnology* 17 (1999): 135–36.

9. Das, "Production of Therapeutic Proteins."

10. Ibid.

11. Yoon, "If It Walks and Moos like a Cow."

12. J. David Bleich, "Kashrut," in *Contemporary Halachic Problems* (New York: Ktav Publishing House, 1989), 3:68.

13. Ibid., 69.

14. Ibid., 70.

15. Talmud *Yoma* 67b; also see "Mitzvah" in *The New Standard Jewish Encyclopedia,* 7th ed., ed. G. Wigoder (New York: Facts on File, 1992), 664.

16. Mishnah *Kilayim* 9:8.

17. Ibid., 1:6.

18. Ibid., 8:1.

19. Shabtai Rappaport, "Genetic Engineering: Technology, Creation and Interference," *ASSIA—Journal of Jewish Medical Ethics and Halacha* 3.1 (1997): 3–4.

20. Shammai Engelmayer, "Learning to Live with Genetic Engineering," *The New Jersey Jewish Standard,* July 31, 1998, p. 6.

21. Nachmanides' commentary on Leviticus 19:19.

22. Walter Jacobs, Central Conference of American Rabbis (CCAR) Responsa 154: "Jewish Involvement in Genetic Engineering," 1992, in *Program Guide X: Cloning,* ed. Union of American Hebrew Congregations Committee on Bio-Ethics (Philadelphia: UAHC, Summer 1998) pp. 61–66.

23. Rappaport, "Genetic Engineering."

24. Ibid.

25. Englemeyer, "Learning to Live with Genetic Engineering."

26. Avraham Steinberg, "Rabbi Shlomo Zalman Auerbach," in *Pioneers in Jewish Medical Ethics,* ed. Fred Rosner (Northvale, N.J.: Jason Aronson, 1997), 112.

27. Mishnah *Kilayim* 9:1.

28. Philip Blackman, ed. *The Mishnah.* Footnote on *Kilayim* 9:1, Mishnayoth vol. 1: Order *Zeraim* (Gateshead, England: Judaica Press, 1964).

29. Avram I. Reisner, "Curiouser and Curiouser: *Teshuvah* on Genetic Engineering," *Conservative Judaism.* 52.3 (2000): 59–72. Note that the *Tosefta* (literally "Supplement"), usually attributed to third-century rabbis Rabbah and Oshaya, is considered analogous to the Mishnah. It contains a collection of laws plus supplemental material. See A. Cohen, *Everyman's Talmud* (New York: E. P. Dutton & Co., 1949), xxx.

30. *Shulhan Arukh, Yoreh Deah* 295:7.

31. Jacobs, CCAR Responsa 154.

32. Moses Maimonides, *The Guide for the Perplexed* (New York: Dover Publications, 1956; 1881 translation by M. Friedlander) bk. 3, Ch. 17, pp. 287–88.

33. *DSM IV, Diagnostic and Statistical Manual of Mental Disorders,* 4th ed. (Washington, D.C.: American Psychiatric Association, 1994), 86–90.

34. A. Arluke, J. Levin, and C. Luke, "The Relationship of Animal Abuse to Violence and Other Forms of Antisocial Behavior," *Journal of Interpersonal Violence* 14 (1999): 963–75.

35. Barbara Wickens, "Seeing Pet Abuse as a Warning," *Maclean's* 3.43, October 26, 1998, p. 72.

36. J. David Bleich, "Animal Experimentation," in *Contemporary Halakhic Problems* (New York: Ktav Publishing House, 1989), 3:209.

37. Ibid., 225.

38. Jacobs, CCAR Responsa 154.

39. Bleich, "Animal Experimentation," 235.

40. Nisson Shulman, "Genetic Issues," in *Jewish Answers to Medical Ethics Questions* (Northvale, N.J.: Jason Aronson, 1998), 13.

41. Reisner, "Curiouser and Curiouser."

42. Nachmanides, commentary on Gen. 1:28.

43. Nachmanides on Leviticus 19:19, as described by Fred Rosner, "Genetic Screening, Genetic Therapy and Cloning in Judaism," *B'Or Ha'torah* 12E (2001): 17–29.

44. Jacobs, CCAR Responsa 154.

45. Bleich, "Animal Experimentation," 71.

46. Ibid., 72.

47. Ibid.

48. Ask Aish #25-2000, "Genome Project." Web URL: <aish.com/rabbi/ATR_viewLinks.asp>.

49. Talmud Hullin 80a.

50. A. Cohen, ed., *The Soncino Chumash* (London: Soncino Press, 1947).

51. N. Scherman and M. Zlotowitz, *The Stone Edition Tanach* (New York: ArtScroll/Mesorah Publications, Ltd, 1996).

52. Binyomin Forst, *The Laws of Kashrut* (New York: Artscroll Mesorah Publications, 1993), 34.

53. Bleich, "Animal Experimentation," 74.

54. Reisner, "Curiouser and Curiouser."

55. Ibid.

56. Ibid.

57. Paul DeKruif, *Microbe Hunters* (New York: Harcourt, Brace and Co., 1926).

58. Rabbi Yechiel Epstein, *Arokh HaShulhan: Yoreh Deah* 84.36, as cited in Reisner, "Curiouser and Curiouser."

59. Forst, *The Laws of Kashrut,* 80–81.

60. Shulman, "Genetic Issues," 18.

61. Tzvee Zahavy, quoted in M. Wahrman, "Treif Tomatoes," *The New Jersey Jewish Standard,* July 31, 1998, p. 6.

62. Ibid.

63. Judith Leff, "The Modern Food Industry and Kashrut," *Jewish Action* (Summer 1994): 39–43.

64. "Seeds of Change," *Consumer Reports* (September 1999): 41–46.

65. Reisner, "Curiouser and Curiouser."

66. Fred Rosner, "Organ Transplantation in Jewish Law," in *Jewish Bioethics,* ed. F. Rosner and J. D. Bleich (New York: Hebrew Publishing Company, 1979), 358–74.

67. Talmud *Sotah* 43b.

68. Azriel Rosenfeld, "Judaism and Gene Design," in *Jewish Bioethics,* ed. Rosner and Bleich, 401–8.

69. Shulman, "Genetic Issues," 17.

70. Reisner, "Curiouser and Curiouser."

71. Shulman, "Genetic Issues," 16–17.

72. Bleich, "Animal Experimentation," 67.

73. Ibid.

74. Ibid.

Chapter 12. Treife *Tomatoes: Brave New Plants (pages 209-26)*

1. David Ow, Keith Wood, and Marlene DeLuca, "Transient and Stable Expression of the Firefly Luciferase Gene in Plant Cells and Transgenic Plants,"

Science 234 (1986): 856–59; Sharon Begley, "Brave New Monkey," *Newsweek*, January 22, 2001, p. 50.

2. Charles Beck and Thomas Ulrich, "Biotechnology in the Food Industry," *BIO/technology* 11 (1993): 895–902.

3. Ibid.

4. Bert Popping, "Methods for the Detection of Genetically Modfied Organisms: Precision, Pitfalls and Proficiency," *American Laboratory* 33(2001): 70–80.

5. "Seeds of Change," *Consumer Reports* (September 1999): 41–46.

6. Sunday Q & A, *The New York Times*, May 13, 2001, p. 25.

7. Jon Palfreman, "Harvest of Fear," *Frontline/Nova*, WGBH Educational Foundation Production, Boston. Original airdate: April 23, 2001.

8. Mishnah *Kilayim* 1:4.

9. Ibid. 1:7.

10. Robert Schery, "Buying Lawn Seed," in *Home Lawn Handbook* (Brooklyn, N.Y.: Brooklyn Botanic Garden, 1988).

11. Mishnah *Kilayim* 2:1.

12. Avram I. Reisner, "Curiouser and Curiouser: *Teshuvah* on Genetic Engineering," *Conservative Judaism* 52.3 (2000): 59–72; regarding *Tosefta*, see ch. 11, n.29.

13. Jill H. Coplan, "But Will the Rabbis Still Eat It?" *The Jerusalem Report*, February 15, 1999, pp. 33–34.

14. Palfreman, "Harvest of Fear."

15. Michael Pollan, "Playing God in the Garden," *The New York Times Magazine*, October 25, 1998, p. 44.

16. Joshua Finkelstein, "Genetically Modified Food," lecture delivered at Symposium on the Ethical Issues of Biogenetic Research, Gerrard Berman Day School Solomon Schechter of New Jersey, Oakland, N.J., February 13, 2001.

17. Nisson Shulman, "Genetic Issues," in *Jewish Answers to Medical Ethics Questions* (Northvale, N.J.: Jason Aronson, 1998), 18–19.

18. Fred Rosner, "Genetic Screening, Genetic Therapy and Cloning in Judaism," *B'Or Ha'torah* 12E (2001): 17–29.

19. Ibid.

20. Avraham Steinberg, "Medical-Halachic Decisions of Rabbi Shlomo Zalman Auerbach (1910–1995)," *ASSIA: Journal of Jewish Medical Ethics and Halacha* 3.1 (1997): 30–43.

21. Finkelstein, "Genetically Modified Foods."

22. Shammai Engelmayer, "Learning "Learning to Live with Genetic Engineering," *The New Jersey Jewish Standard*, July 31, 1998, p. 6.

23. Ibid. Also see n.18.

24. Ibid.

25. Fred Rosner, "Genetic Engineering," in *Jewish Bioethics,* ed. F. Rosner and J. D. Bleich (New York: Hebrew Publishing Company, 1979), 409–20.

26. Judith Leff, "The Modern Food Industry and Kashrut," *Jewish Action* (Summer 1994): 39–43.

27. Ibid.

28. Ibid.

29. Ibid.

30. Judith Leff, "Kashrus in the Year 2000," *The Jewish Observer* (September 2000), 35–38.

31. Kathryn Brown, "Seeds of Concern," *Scientific American* (April 2001): 52–57.

32. Finkelstein, "Genetically Modified Foods."

33. Immanuel Jakobovits, *Jewish Medical Ethics* (New York: Bloch Pub., 1975), 264–66.

34. Shulman, "Genetic Issues," 13–15.

35. Shabtai Rappaport, "Genetic Engineering: Technology, Creation and Interference," *ASSIA: Journal of Jewish Medical Ethics and Halacha* 3.1 (1997): 3–4.

36. Robert Gordis, "Ecology and the Judaic Tradition," *Judaic Ethics for a Lawless World* (New York: Jewish Theological Seminary of America, 1986), 118.

37. Ibid.

Chapter 13. When Science and Scripture Collide (pages 227-37)

1. President Bill Clinton, quoted in Herb Keinon, "Jews and the Genome," *The Jerusalem Post* (Internet edition), July 13, 2000. Website URL: <www.jpost .com/Editions/2000/07/06/Features/Features.9295.html>.

2. Talmud *Niddah* 31a.

3. Talmud *Niddah* 16b.

4. Rashi's commentary on Numbers 24:3.

5. Rashi's commentary on Leviticus 19:36.

6. Michael Drosnin, *The Bible Code* (New York: Simon and Schuster, 1997).

7. Philip Hilts, "Decipherers of Dead Sea Scrolls Turn to DNA Analysis for Help," *The New York Times,* March 28, 1995, p. C1.

8. Ibid.

9. Maimonides' commentary on Mishnah *Sanhedrin* 10:1.

10. Bracha Etengoff, "Biotechnology and the Resurrection," *Derech HaTeva: Torah and Science Journal of Stern College for Women, Yeshiva University* 3 (1998–99/5759): 25–27.

11. International Human Genome Sequencing Consortium, "Initial Sequencing and Analysis of the Humane Genome," *Nature* 409 (2001): 860–921.

12. Gerald Rubin, "Comparing Species," *Nature* 409 (2001): 820–21.

13. Fred Rosner, "The Physician's Prayer Attributed to Moses Maimonides," in *The Medical Legacy of Moses Maimonides* (Hoboken, N.J.: Ktav Publishing House, 1998), 273–90.

14. *Oath of Maimonides,* Jewish Virtual Library, American-Israeli Cooperative Enterprise, 2001. Website URL: <www.us-israel.org/jsource/Judaism/oath .html>.

15. Ibid.

16. Dr. Francis Collins quoted in Herb Keinon, "Jews and the Genome," *The Jerusalem Post* (Internet edition), July 13, 2000.

Selected Bibliography

The selected sources listed below are recommended for those who would like to delve deeper into some of the issues addressed by this book. They include research articles in biology, biotechnology, and genetics; articles by Judaic scholars; books on bioethics, Jewish bioethics, and other scientific sources and modern Jewish commentaries. Since the topics addressed here are cutting edge, sources include newspapers, writings on the Internet, and film documentaries.

Biblical and Talmudic Sources

Resources for biblical and talmudic study on levels ranging from beginner through advanced scholar abound, and many resources have been translated into English. This book draws on sources from the *Torah, Neveim,* and *Ketuvim* (Hebrew acronym, *Tanakh*), which includes the Bible, the Prophets, and the Writings of the Old Testament.

In addition to the comprehensive *Mikraot Gedolot* edition of the Torah, which includes Rashi, Ramban, Daat Zekeinim, Onkelos, and other commentaries (in Hebrew and Aramaic), several English translations of **Torah and Tanakh** are available, such as:

Ben Isaiah, Abraham and Benjamin Sharfman, translators. *The Pentateuch and Rashi's Commentary: A Linear Translation into English.* Brooklyn, N.Y.: S. S. & R. Publishing Co., 1949.

Cohen, A., editor. *The Soncino Chumash.* London: Soncino Press, 1947.

Elman, Yaakov, translator. *The Living Nach.* New York: Moznaim Publishing Corp, 1994.

Harkavy, Alexander, translator. *The Twenty-Four Books of the Holy Scriptures.* (New King James version.) New York: Hebrew Publishing Company, 1916.

Hertz, J. H., editor. *The Pentateuch and Haftorahs.* London: Soncino Press, 1960.

Plaut, W. Gunther, and Bernard Bamberger, commentators. *The Torah, A Modern Commentary.* New York: Union of American Hebrew Congregations, 1981.

Scherman, N., and M. Zlotowitz, editors. *The Stone Edition Chumash.* New York: ArtScroll/Mesorah Publications, Ltd, 1993.

———. *The Stone Edition Tanach.* New York: ArtScroll/Mesorah Publications, Ltd, 1996.

Translations of the **Mishnah** include:

Blackman, Philip. *Mishnayoth.* Second edition. Gateshead, England: Judaica Press, 1990.

Danby, Herbert, translator. *The Mishnah*. London: Oxford University Press, 1933.

Kehati, Pinhas. *The Mishnah*. Translated by Edward Levin. Jerusalem: Eliner Library, 1994.

Most talmudic citations included in this book refer to passages from the **Babylonian Talmud** *(Talmud Bavli)*. Translations of the Gemara are available in the following editions:

The Schottenstein Edition. *Talmud Bavli*. New York: ArtScroll Publications, 1990–.

The Soncino Press, *Hebrew-English Edition of the Babylonian Talmud*, London: 1990.

The Soncino Talmud on CD-ROM, Davka Corporation, Chicago, Ill. (Hebrew/Aramaic text of Talmud with Rashi's commentary and English translation of Mishnah and Talmud.) 1991–1995, Institute for Computers in Jewish Life, Danka Corp., and Judaica Press.

The Talmud—The Steinsaltz Edition. Commentary by Adin Steinsaltz. New York: Random House, 1990–.

Other commentaries that are available in English include:

Maimonides, Moses. *The Guide for the Perplexed*. Translated in 1881 by M. Friedlander. New York: Dover Publications, 1956.

Maimonides, Moses. *Mishnah Torah*. Edited and translated by Philip Birnbaum. New York: Hebrew Publishing Co., 1944.

Biological Sources

In addition to the sources listed below, there are dozens of introductory textbooks in biology and genetics that provide background information on many of the topics covered in this book. Additional books are readily available on topics such as cloning, genetic engineering, genetically modified foods, and genetic disorders.

BOOKS

Amelar, Richard, Lawrence Dubin, and Patrick Walsh. *Male Infertility*. Philadelphia: W. B. Saunders Co., 1977.

Beauchamp, Tom, and James F. Childress. *Principles of Biomedical Ethics*. New York: Oxford University Press, 1979.

Beauchamp, Tom, and LeRoy Walters. *Contemporary Issues in Bioethics*. Encino, Calif.: Dickenson Pub. Co., 1978.

Bial, Morrison David. *Liberal Judaism at Home*. New York: Union of American Hebrew Congregations, 1971.

Bleich, J. David. *Bioethical Dilemmas*. Hoboken, N.J.: Ktav Publishing House, 1998.

———. *Contemporary Halakhic Problems*. Vol 1. New York: Ktav Publishing House and Yeshiva University Press, 1977.

———. *Contemporary Halakhic Problems.* Vol. 3. New York: Ktav Publishing House, 1989.

———. *Judaism and Healing: Halakhic Perspectives.* Hoboken, N.J.: Ktav Publishing House, 1981.

Branover, Herman, and Ilana Coven Attia, editors. *Science in the Light of Torah.* Northvale, N.J.: Jason Aronson, 1994.

Cohen, Abraham editor. *Everyman's Talmud.* New York: E. P. Dutton and Co., 1949.

Cohen, Arthur, and Paul Mendes-Flohr, editors. *Contemporary Jewish Religious Thought.* New York: Charles Scribner's Sons, 1987.

DeKruif, Paul. *Microbe Hunters.* New York: Harcourt, Brace and Co., 1926.

Dorff, Elliot N. *Matters of Life and Death.* Philadelphia and Jerusalem: Jewish Publication Society, 1998.

Drosnin, Michael. *The Bible Code.* New York: Simon and Schuster, 1997.

DSM IV, Diagnostic and Statistical Manual of Mental Disorders. Fourth edition. Washington, D.C.: American Psychiatric Association, 1994.

Feldman, David M. *Birth Control in Jewish Law.* New York: New York University Press, 1968; third edition, 1995.

Feldman, Emanuel, and Joel B. Wolowelsky, editors. *Jewish Law and the New Reproductive Technologies.* Hoboken, N.J.: Ktav Publishing House, 1997.

Fendel, Zechariah. *Anvil of Sinai.* New York: Hashkafah Pubs., 1977.

Forst, Binyomin. *The Laws of Kashrut.* New York: ArtScroll Mesorah Publications, 1993.

Gordis, Robert. *Judaic Ethics for a Lawless World.* New York: Jewish Theological Seminary of America, 1986. Pp. 113–22.

Grazi, Richard V., editor. *Be Fruitful and Multiply: Fertility Therapy and the Jewish Tradition.* Jerusalem: Genesis Jerusalem Press, 1994.

Hess, Beth, Elizabeth W. Markson, and Peter J. Stein. *Sociology.* Fifth edition. Needham Heights, Mass.: Allyn and Bacon, 1996.

Huxley, Aldous. *Brave New World.* New York, London: Harper and Brothers, 1932.

Jacobs, Lewis. *What Does Judaism Say About . . . ?* New York: Quadrangle/The New York Times Book Co., 1973.

Jakobovits, Immanuel. *Jewish Medical Ethics.* New York: Bloch Publishing Co, 1959; second edition, 1975.

Jansen, Robert. *Overcoming Infertility.* New York: W. H. Freeman, 1997.

Jarashow, Jonathan. *The Silent Psalms of Our Son.* Nanuet, N.Y.: Philip Feldheim, 2001.

Lewis, Ricki. *Human Genetics: Concepts and Applications.* Fourth edition. New York: McGraw-Hill, 2001.

Mappes, Thomas A., and David DeGrazia. *Biomedical Ethics.* New York: McGraw-Hill, 1996.

Meyer, Michael A. *Response to Modernity.* New York: Oxford University Press, 1988.

Parfitt, Tudor. *Journey to the Vanished City.* New York: St. Martin's Press, 1992.

Pellegrino, Edmund, and Alan Faden, editors. *Jewish and Catholic Bioethics.* Washington, D.C.: Georgetown University Press, 1999.

Preuss, Julius. *Biblical and Talmudic Medicine*. Translated and edited by Fred Rosner. Northvale, N.J.: Jason Aronson, 1993; first published in 1911.

Ridley, Aaron. *Beginning Bioethics*. New York: St. Martin's Press, 1998.

Ridley, Matt. *Genome*. New York: HarperCollins, 1999.

Rosner, Fred. *The Medical Legacy of Moses Maimonides*. Hoboken, N.J.: Ktav Publishing House, 1998.

———. *Medicine and Jewish Law*. Vol 2. Northvale, N.J.: Jason Aronson, 1993.

———. *Modern Medicine and Jewish Ethics*. Hoboken, N.J.: Ktav Publishing House, and New York: Yeshiva University Press, 1986.

———. *Sex Ethics in the Writings of Moses Maimonides*. Northvale, N.J.: Jason Aronson, 1994.

———, editor. *Pioneers in Jewish Medical Ethics*. Northvale, N.J.: Jason Aronson, 1997.

Rosner, Fred, and J. D. Bleich, editors. *Jewish Bioethics*. Brooklyn, N.Y.: Hebrew Publishing Co., 1979. Second edition: Hoboken, N.J.: Ktav Publishing House, 2000.

Ross, James R. *Fragile Branches: Travels Through the Jewish Diaspora*. New York: Riverhead Books, 2000.

Shulman, Nisson E. *Jewish Answers to Medical Ethics Questions*. Northvale, N.J.: Jason Aronson, 1998.

Tendler, Moshe. *Pardes Rimonim: A Manual for the Jewish Family*. Hoboken, N.J.: Ktav Publishing House, 1988.

Tillem, Ivan, editor. *The 1986 Jewish Directory and Almanac*. New York: Pacific Press, 1986.

Union of American Hebrew Congregations Committee on Bio-Ethics. *Program Guide V: Genetic Screening and the Human Genome Project*. Philadelphia: UAHC, Spring 1992.

———. *Program Guide IX: Organ Donations and Transplantation*. Philadelphia: UAHC, Spring 1997.

———. *Program Guide X: Cloning*. Philadelphia: UAHC, Summer 1998.

———. *Program Guide XI: Infertility and Assisted Reproduction*. New York: UAHC, Autumn 1999.

ARTICLES AND OTHER MEDIA

Anderson, S. L., R. Coli, I. W. Daly, E. A. Kichula, M. J. Rork, S. A. Volpi, J. Ekstein, and B. Y. Rubin. "Familial Dysautonomia Is Caused by Mutations of the IKAP Gene." *American Journal of Human Genetics* 68 (2001): 753–58.

Arluke, A., J. Levin, and C. Luke. "The Relationship of Animal Abuse to Violence and Other Forms of Antisocial Behavior." *Journal of Interpersonal Violence* 14 (1999): 963–75.

Baharloo, Siamak, Paul A. Johnston, Susan K. Service, Jane Gitschier, and Nelson B. Freimer. "Absolute Pitch: An Approach for Identification of Genetic and Nongenetic Components." *American Journal of Human Genetics* 62 (1998): 224–31.

Baharloo, Siamak, Susan Service, Neil Risch, Jane Gitschier, and Nelson Freimer. "Familial Aggregation of Absolute Pitch." *American Journal of Human Genetics* 67 (2000): 755–58.

Baltimore, David. "Our Genome Unveiled." *Nature* 409 (2001): 814–16.

Barg, P. E., M. Z. Wahrman, B. E. Talansky, and J. W. Gordon. "Capacitated, Acrosome Reacted but Immotile Sperm, When Microinjected Under the Mouse Zona Pellucida, Will Not Fertilize the Oocyte." *Journal of Experimental Zoology* 237 (1986): 365–74.

Beck, Charles, and Thomas Ulrich. "Biotechnology in the Food Industry." *BIO/technology* 11 (1993): 895–902.

Begley, Sharon. "Brave New Monkey." *Newsweek,* January 22, 2001, p. 50.

———. "Cellular Divide." *Newsweek,* July 9, 2001, p. 22.

Blakeslee, Sandra. "Watching How the Brain Works as It Weighs a Moral Dilemma." *The New York Times,* September 25, 2001, p. F3.

Blank, Barbara Trainin. "The Ethics of Cloning." *Hadassah Magazine* (June/July 1997): 23–24.

Bleich, J. David. "Cloning: Homologous Reproduction and Jewish Law." *Tradition* 32 (1998): 47–86.

———. "In Vitro Fertilization: Questions of Maternal Identity and Conversion." *Tradition* 25 (1991): 82–102.

Brander, Kenneth. "Artificial Insemination and Surrogate Motherhood through the Prism of Jewish Law." *B'Or Ha'Torah* 12E (2001): 59–65.

Breast Cancer Linkage Consortium. "Cancer Risks in BRCA2 Mutation Carriers." *Journal of the National Cancer Institute* 91 (1999): 1310–16.

Breitowitz, Yitzchok A. "Halakhic Approaches to the Resolution of Disputes Concerning the Disposition of Preembryos." *Tradition* 31.1 (1996): 64–91.

Brown, Kathryn. "Seeds of Concern." *Scientific American* (April 2001): 52–57.

Broyde, Michael. "The Cloning." *Emunah* (Spring/Summer 1998): 14–17.

———. "Cloning People: A Jewish Law Analysis of the Issues." *Connecticut Law Review* 30 (1998): 503–35.

———. "Cloning People and Jewish Law: A Preliminary Analysis." *Journal of Halacha and Contemporary Society* 34 (1997): 27–65.

Cann, Rebecca L. "In Search of Eve." *The Sciences* (Sept/Oct 1987): 30–37.

Cavazzana-Calvo, M., S. Hacein-Bey, G. De Saint Basile, F. Gross, E. Yvon, P. Nusbaum, F. Selz, C. Hue, S. Certain, J.-L. Casanova, P. Bousso, F. LeDeist, and A. Fisher. "Gene Therapy of Human Severe Combined Immunodeficiency (SCID)—XI Disease." *Science* 288 (2000): 669–72.

Central Conference of American Rabbis (CCAR). Contemporary American Reform Responsa 20. "Genetic Engineering." February 1978. Website URL: <www.ccarnet.org/cgi-bin/respdisp.pr?me-20&year-carr>.

———. *Reform Responsa* 5757.2. "In Vitro Fertilization and the Status of the Embryo." 1997. Website URL: <http://www.ccarnet.org/resp>.

———. "A Statement of Principles for Reform Judaism." Adopted at the 1999 Pittsburgh Convention, May 26, 1999. Website URL: <http://ccarnet.org/plat forms/principles.html>.

Check, J. H. and D. Katsoff. "A Prospective Study to Evaluate the Efficacy of

Modified Swim-up Preparation for Male Sex Selection." *Human Reproduction* 8 (1993): 211–14.

Chen, Christina. "Population-wide Gene Searches." UCSF News Archive: February 1, 1997. Website URL: <www.ucsf.edu/~adcom/listserv/ucsfnews/0007.html>.

Chen, M. J., H. F. Guv, and E. S. Ho. "Efficiency of Sex-selection of Spermatozoa by Albumin Separation Method Evaluated by Double-labelled Fluorescence In-situ Hybridization." *Human Reproduction* 12 (1997): 1920–26.

Coplan, Jill H. "But Will the Rabbis Still Eat It?" *The Jerusalem Report*, Feb. 15, 1999, pp. 33–34.

Das, Rathin. "Production of Therapeutic Proteins from Transgenic Animals." *American Biotechnology Laboratory* 19.2 (February 2001): 60–64.

DeJonge, C. J., S. P. Flaherty, A. M. Barnes, N. J. Swann, and C. D. Matthews. "Failure of Multitube Sperm Swim-up for Sex Preselection." *Fertility and Sterility* 67 (1997): 1109–14.

Diamond, Jared. "Who Are the Jews?" *Natural History* (November 1993): 12–19.

Dugger, Celia W. "Modern Asia's Anomaly: The Girls Who Don't Get Born." *The New York Times,* May 6, 2001, p. wk. 4.

Effron, Sonni. "Japanese Couples Think Pink." *The Los Angeles Times,* November 15, 1999, p.1.

Elliman, Wendy. "Menashe's Children Come Home." *Hadassah Magazine* (October 1999).

Engelmayer, Shammai. "Learning to Live with Genetic Engineering." *The New Jersey Jewish Standard,* July 31, 1998, p. 6.

Eppstein, L. "UCSF Researchers Seek Subjects for Perfect-Pitch Study." *Jewish Bulletin of Northern California,* July 25, 1997.

Ericsson, R., C. Langevin, and M. Nishino. "Isolation of Fractions Rich in Human Y Sperm." *Nature* 246 (1973): 421–24.

Etengoff, Bracha. "Biotechnology and the Resurrection." *Derech HaTeva: Torah and Science Journal of Stern College for Women, Yeshiva University* 3 (1998–99/5759): 25–27.

Evangelista, Andy, and Jennifer O'Brien. "Researchers Use Web to Aid Musical Gene Hunt." *UCSF Daybreak News,* University of California, February 28, 1998. Website URL: <itssrv1.ucsf.edu/daybreak/1998/02/206_musi.htm>.

Flancbaum, Louis. "Using Jewish Medical Ethics to Appreciate the Relative Among the Absolute." *B'Or Ha'Torah* 12E (2001): 95–104.

Frimer, Dov I. 1989. "Establishing Paternity by Means of Blood Type Testing." *ASSIA: Journal of Jewish Medical Ethics and Halacha* 1.2 (1989): 20–35.

Fugger, E. F., S. H. Black, K. Keyvanfar, J. D. Schulman. "Births of Normal Daughters after MicroSort Sperm Separation and Intrauterine Insemination, In Vitro Fertilization, or Intracytoplasmic Sperm Injection." *Human Reproduction* 13 (1998): 2367–70.

Garber, Judy. "A 40-Year-Old Woman with a Strong Family History of Breast Cancer." *JAMA* 282 (1999): 1953–60.

Gopan, Sata, and Kenneth Offit. "Lifetime Risks of Breast Cancer in Ashkenazi

Jewish Carriers of BRCA1 and BRCA2 Mutations." *Cancer Epidemiology Biomarkers and Prevention* 10 (2001): 467–73.

Grodin, Michael A. "Halakhic Dilemmas in Modern Medicine." *The Journal of Clinical Ethics* 6.3 (1995): 218–21.

Halperin, Mordechai. "Human Genome Mapping: A Jewish Perspective." *ASSIA: Journal of Jewish Medical Ethics and Halacha* 3.2 (1998): 30–33.

Hammer, M. F., A. J. Redd, E. T. Wood, M. R. Bonner, H. Jarjanazi, T. Karafet, S. Santachiara-Benerecetti, A. Oppenheim, M. A. Jobling, T. Jenkins, H. Ostrer, and B. Bonne-Tamir. "Jewish and Middle Eastern Non-Jewish Populations Share a Common Pool of Y-Chromosome Biallelic Haplotypes." *Proceedings of the National Academy of Science* 97 (2000): 6769–74.

Hilts, Philip. "Decipherers of Dead Sea Scrolls Turn to DNA Analysis for Help." *The New York Times,* March 28, 1995, p. C1.

International Human Genome Sequencing Consortium. "Initial Sequencing and Analysis of the Human Genome." *Nature* 409 (2001): 860–21.

Johns Hopkins University School of Medicine, Epidemiology-Genetics Program in Psychiatry, Winter 1999 Newsletter. Website URL: <www.med.jhu.edu/epigen/winter_1999.htm>.

Jost, B., J.-L. Vilotte, I. Duluc, J.-L. Rodeau, J.-N. Freund. "Production of Low-Lactose Milk by Ectopic Expression of Intestinal Lactase in the Mouse Mammary Gland." *Nature Biotechnology* 17 (1999): 160–64.

Kay, M. A., C. S. Manno, M. V. Ragni, P. J. Larson, L. B. Couto, A. McClelland, B. Glader, A. J. Chew, S. J. Tai, R. W. Herzog, V. Arruda, F. Johnson, C. Scallan, E. Skarsgard, A. W. Flake, and K. A. High. "Evidence for Gene Transfer and Expression of Factor IX in Haemophilia B Patients Treated with an AAV Vector." *Nature Genetics* 24 (2000): 257–61.

Keinon, Herb. "Jews and the Genome." *The Jerusalem Post* (Internet edition), July 13, 2000.

Kleiman, Yaakov. "The DNA Chain of Tradition." *Jewish Action* (Winter 5760/1999).

Kolata, Gina. "Fertility Ethics Authority Approves Sex Selection." *The New York Times,* September 28, 2001, p. A 16.

———. "Researchers Find Big Risk of Defect in Cloning Animals." *The New York Times,* March 25, 2001, p. 1.

———. "Researchers Say Embryos in Labs Aren't Available." *The New York Times,* August 26, 2001, p. 1.

Kukin, Marrick. "Tay Sachs and the Abortion Controversy." *Journal of Religion and Health* 20.3 (1981): 224–42.

Lancet. "Overexcitement on Embryo Stem Cells," editorial, 356. 9231, August 26, 2000.

Langston, A. A., K. E. Malone, J. D. Thompson, J. R. Daling, E. A. Ostrander. "BRCA1 Mutations in a Population-Based Sample of Young Women with Breast Cancer." *New England Journal of Medicine* 334 (1996): 137–42.

Lazarus, Janine. "At the Jewish Doorstep in Africa." *Hadassah Magazine* (January 2001).

Lazarus, Moshe, Reuven Subar, and Avrohom Lefkowitz. "Ask the Rabbi." In Ohr Somayach Institutions, Jerusalem, Israel, November 12, 1994, Issue 42. Website URL: <http://ohr.org.il/ask/ask042.htm>.

Leff, Judith. "Kashrus in the Year 2000." *The Jewish Observer* (September 2000): 35–38.

———. "The Modern Food Industry and Kashrut." *Jewish Action* (Summer 1994): 39–43.

Levinson, Gabe. "Remote Destinations." *The Jewish Week,* October 27, 2000, p. 43.

Levinson, Jay. "An Halakhic Reconsideration of Victim Identification." *ASSIA: Journal of Jewish Medical Ethics and Halacha* 4.1 (February 2001): 55–57.

———. "Halachic Authority Limits Use of DNA Evidence for Identification." *The Jewish Press,* February 23, 2001, p. 10.

Liu, P., and G. A. Rose. "Social Aspects of More than 800 Couples Coming Forward for Gender Selection of Their Children." *Human Reproduction* 10.4 (1995): 968–71.

———. "Sex Selection: The Right Way Forward." *Human Reproduction* 11.11 (1996): 2343–45.

Loike, John D., and Avraham Steinberg. "Human Cloning and Halakhic Perspectives." *Tradition* 32 (1998): 31–46.

Lucotte, G., and P. Smets. "Origins of Falasha Jews Studied by Haplotypes of the Y Chromosome." *Human Biology* 71 (1999): 989–93.

McLaughlin, Katy. "Confronting Disaster Halachically." *Jewish Renaissance Media,* September 25, 2001. Website URL: <ww.jewish.com/terror/disaster 0925.shtml>.

Mosenkis, Ari. "Genetic Screening for Breast Cancer Susceptibility: A Torah Perspective." *Journal of Halacha and Contemporary Society* 34 (1997): 5–26.

Mudur, Ganapati. "Indian Medical Authorities Act on Antenatal Sex Selection." *British Medical Journal* 319 (1999): 401.

National Bioethics Advisory Commission. *Ethical Issues in Human Stem Cell Research, Volume I: Report and Recommendation of the National Bioethics Advisory Commission.* Rockville, Md.: National Bioethics Advisory Commission, 1999.

Neergaard, Lauran. "Ethicist Approves of Some Sex Selection." *The Bergen Record,* September 29, 2001, p. A 12.

Nimoy, Leonard. "So Human." *Reform Judaism Magazine* (August 1999).

Ow, David, Keith Wood, and Marlene DeLuca. "Transient and Stable Expression of the Firefly Luciferase Gene in Plant Cells and Transgenic Plants." *Science* 234 (1986): 856–59.

Palfreman, Jon. "Harvest of Fear." Frontline/Nova, WGBH Educational Foundation Production, Boston. Original airdate: April 23, 2001.

Palmiter, R. D., R. L. Brinster, R. E. Hammer, M. E. Trumbauer, M. G. Rosenfield, N. C. Birnberg, and R. M. Evans. "Dramatic Growth of Mice That Develop from Eggs Microinjected with Metallothionein-Growth Hormone Fusion Genes." *Nature* 300 (1982): 611–15.

Pennings, G. "Family Balancing as a Morally Acceptable Application of Sex Selection." *Human Reproduction* 11.11 (1996): 2339–43.

Pollan, Michael. "Playing God in the Garden." *The New York Times Magazine,* October 25, 1998, p. 44.

Popping, Bert. "Methods for the Detection of Genetically Modified Organisms: Precision, Pitfalls and Proficiency." *American Laboratory* 33 (2001): 70–80.

Powledge, T. M., and M. Rose. "The Great DNA Hunt." *Archeology* (September/October 1996): 36–44.

Ramachandran, R. "In India, Sex Selection Gets Easier." *The UNESCO Courier* 52 (1999): 29.

Ramsey, Sarah. "UK Government Looks to Expand Research on Embryos." *The Lancet* 356.9244, November 25, 2000.

Rappaport, Shabtai. "Genetic Engineering: Technology, Creation and Interference." *ASSIA: Journal of Jewish Medical Ethics and Halacha* 3.1 (1997): 3–4.

Regalado, Antonio, Jill Carroll, and Laura Johannes. "Stem-Cell Scrum: Do 60 Lines Even Exist?" *The Wall Street Journal,* August 14, 2001, p. 16.

Reisner, Avram I. "Curiouser and Curiouser: *Teshuvah* on Genetic Engineering." *Conservative Judaism* 52.3 (2000): 59–72.

Reuters Medical News. "European Law May Change Regarding Stem Cell Research." July 10, 2000. Website URL: <www.medscape.com/reuters/prof/2000/07/07.10/20000710ethc002.html>.

———. "Italian Experts Back Human Stem Cell Cloning." December 28, 2000. Website URL: <www.medscape.com/reuters/prof/2000/12/12.29/20001228 ethc001.html>.

Richards, W. E., S. M. Dobin, V. Malone, A. B. Knight, and T. J. Kuehl. "Evaluating Sex Chromosome Content of Sorted Human Sperm Samples with Use of Dual-color Fluorescence In Situ Hybridization." *American Journal of Obstetrics and Gynecology* 176 (1997): 1172–78.

Ritte, U., E. Neufeld, M. Broit, D. Shavit, and U. Motro. "The Differences Among Jewish Communities—Maternal and Paternal Contributions." *Journal of Molecular Evolution* 37 (1993): 435–40.

Rose, G. A., and A. Wong. "Experiences in Hong Kong with the Theory and Practice of the Albumin Column Method of Sperm Separation for Sex Selection." *Human Reproduction* 13 (1998): 146–49.

Rosenfeld, A. "Judaism and Gene Design." *Tradition* 13 (1972): 71–80.

Rosner, Fred. "Genetic Screening, Genetic Therapy and Cloning in Judaism." *B'Or Ha'torah* 12E (2001): 17–29.

———. "Recombinant DNA, Cloning and Genetic Engineering in Judaism." *N.Y. State Journal of Medicine* 79 (1979): 1442.

Rubin, Gerald. "Comparing Species." *Nature* 409 (2001): 820–21.

Samber, Sharon. "For Jews Debate Remains in Embryonic Stage." *New Jersey Jewish Standard,* July 27, 2001, p. 16.

Samura, O., N. Mihara, H. He, E. Okamoto, and K. Ohama. "Assessment of Sex Chromosome Ratio and Aneuploidy Rate in Motile Spermatozoa Selected by Three Different Methods." *Human Reproduction* 12 (1997): 2437–42.

"Seeds of Change." *Consumer Reports* (September 1999): 41–46.

Shamblott, Michael, Joyce Axelman, Shunping Wang, Elizabeth M. Bugg, John W. Littlefield, Peter J. Donovan, Paul D. Blumenthal, George R. Huggins, and John D. Gearhart. "Derivation of Pluripotent Stem Cells from Cultured Human Primordial Germ Cells." *Proceedings of The National Academy of Science.* 95.23 (1998): 13726–31.

Shushan, A., and J. G. Schenker. "Prenatal Sex Determination and Selection." *Human Reproduction* 8 (1993): 1545–49.

Siegel, Judy. "Potential Cure for Heart Disease, Diabetes." *Jerusalem Post,* August 2, 2001, p. 1.

Siegel-Itzkovich, Judy. "How a Hole in the Head Made Medical History." *The Jerusalem Post* (Internet Ed.), October 15, 2001. Website URL: <www.jpost.com/Editions/2001/10/14/Health>.

Sills, E. S., D. Goldschlag, D. P. Levy, O. K. Davis, Z. Rosenwaks. "Preimplantation Genetic Diagnosis: Considerations for Use in Elective Human Embryo Sex Selection." *Journal of Assisted Reproduction and Genetics* 16 (1999): 509–11.

Skorecki, Karl, Sara Selig, Shraga Blazer, Robert Bradman, Neil Bradman, P. J. Waburton, Monica Imajlowicz, and Michael Hammer. "Y Chromosomes of Jewish Priests." *Nature* 385 (1997): 32.

Slaugenhaupt, S. A., A. Blumenfeld, S. P. Gill, M. Leyne, J. Mull, M. P. Cuajungco, C. B. Liebert, B. Chadwick, M. Idelson, L. Reznik, C. M. Robbins, I. Makalowska, M.J. Brownstein, D. Krappmann, C. Scheidereit, C. Maayan, F. B. Axelrod, and J. F. Gusella. "Tissue-specific Expression of a Splicing Mutation in the IKBKAP Gene Causes Familial Dysautonomia." *American Journal of Human Genetics* 68 (2001): 598–605.

Slotnick, R. N., and J. E. Ortega. "Monoamniotic Twinning and Zona Manipulation: A Survey of U.S. IVF Centers Correlating Zona Manipulation Procedures." *Journal of Assisted Reproduction and Genetics* 13.5 (1996): 381–85.

Steinberg, Avraham. "Medical-Halachic Decisions of Rabbi Shlomo Zalman Auerbach (1910–1995)." *ASSIA: Journal of Jewish Medical Ethics and Halacha* 3.1 (1997): 30–43.

Steinberg, A., and J. D. Loike. "Human Cloning: Scientific, Ethical and Jewish Perspectives." *ASSIA: Journal of Jewish Medical Ethics and Halacha* 3 (1998): 11–19.

Stolberg, Sheryl Gay. "Scientists Create Scores of Embryos to Harvest Cells." *The New York Times,* July 11, 2001, p. 1.

Struewing, J. P., D. Abeliovich, and T. Peretz. "The Carrier Frequency of the BRCA1 185delAG Mutation Is Approximately 1 Percent in Ashkenazi Jewish Individuals." *Nature Genetics* 11 (1995): 198–200.

Szabo, C. I., and M. C. King. "Inherited Breast and Ovarian Cancer." *Human Molecular Genetics* 4 (1995): 1811–17.

Talbot, Margaret. "Desire to Duplicate." *The New York Times Magazine,* February 4, 2001, p. 41.

Tendler, Moshe D. "Cell and Organ Transplantation: The Torah Perspective." *B'Or Ha'Torah* 12E (2001): 31–40.

Thomas, Mark G., Tudor Parfitt, Deborah A. Weiss, Karl Skorecki, James F. Wilson, Magdel le Roux, Neil Bradman, and David B. Goldstein. "Y Chromosomes Traveling South: The Cohen Modal Haplotype and the Origins of the Lemba—the 'Black Jews of Southern Africa.'" *American Journal of Human Genetics* 66 (2000): 674–86.

Thomas, Mark G., Karl Skorecki, Haim Ben-Ami, Tudor Parfitt, Neil Bradman, and David Goldstein. "Origins of Old Testament Priests." *Nature* 394 (1998): 138–140.

Thomson, James, Joseph Itskovitz-Eldor, Sander S. Shapiro, Michelle A. Waknitz, Jennifer J. Swiergiel, Vivienne S. Marshall, and Jeffrey M. Jones. "Embryonic Stem Cell Lines Derived from Human Blatocysts." *Science* 282.5391 (1998): 1145–47.

Wade, Nicholas. "DNA Backs a Tribe's Tradition of Early Descent from the Jews." *The New York Times,* May 9, 1999, p. 1.

———. "Scientists Report Two Major Advances in Stem-Cell Research." *The New York Times,* April 27, 2001.

Wadman, Meredith. "So You Want a Girl?" *Fortune* 143.4, February 19, 2001, pp. 174–82.

Wahrman, Miryam. "Too Many Choices?" *The Jewish Standard,* October 16, 1998, p. 8.

———. "Treif Tomatoes." *The New Jersey Jewish Standard,* July 31, 1998, p. 6.

Wahrman, Miryam, J. Victor Reyniak, Andrea Dunaif, Debra Sperling, and Jon Gordon. "Human Egg Pathology: Oocyte Recovery and Egg Morphology as Related to Patient Diagnosis, Fertilization Rate, and Early Development." *J. Gynecological Endocrinology* 1 (1985): 12–19.

Weiss, Rick. "Human Genetic Code Complex in Its Simplicity." *The Bergen Record,* February 11, 2001.

Whitelaw, Bruce. "Toward Designer Milk." *Nature Biotechnology* 17 (1999): 135–36.

Wickens, Barbara. "Seeing Pet Abuse as a Warning." *Maclean's* 3.43, October 26, 1998, p. 72.

Wilmut, I., A. E. Schneike, J. McWhir, A. J. Kind, and K. H. S. Campbell. "Viable Offspring Derived from Fetal and Adult Mammalian Cells." *Nature* 385 (1997): 810–13.

Winerip, M. "Fighting for Jacob." *The New York Times Magazine,* December 6, 1998.

Wooster, R, G. Gignell, J. Lancaster, S. Swift, and S. Deal. "Identification of the Breast Cancer Susceptibility Gene BRCA2." *Nature* 378 (1995): 789–99.

Yoon, Carol K. "If It Walks and Moos Like a Cow, It's a Pharmaceutical Factory." *The New York Times,* May 1, 2000, p. A20.

Index